W9-BUM-467

EXTRAORDINARY GOATS

Meetings with Remarkable Goats, Caprine Wonders & Horned Troublemakers

Janet Hurst • Dennis Pernu
Darwin Holmstrom • Brad Kessler
Elizabeth Noll • Steve Roth

Voyageur Press

Editor: Elizabeth Noll

Art Director: Cindy Samargia Laun

Cover Design: Gavin Duffy

Interior Design: Ryan Scheife

Layout: Diana Boger

On the front cover (left to right): Julia Remezova/Shutterstock, © dbimages / Alamy, Sean Gallup/Getty Images

On the back cover (clockwise from top): Trinity Mirror/Mirrorpix/Alamy, public domain, INTERFOTO/ Alamy, Janet Hurst, public domain, Faiz Balabil/Alamy, Lebrecht Music and Arts Photo Library/Alamy, The Walters Art Museum, makspogonii/Shutterstock, public domain (last two)

First published in 2014 by Voyageur Press, an imprint of Quarto Publishing Group USA Inc., 400 First Avenue North, Suite 400, Minneapolis, MN 55401 USA

ISBN-13: 978-0-7603-4565-8

Library of Congress Cataloging-in-Publication Data

Extraordinary goats : meetings with remarkable goats, caprine wonders, & horned troublemakers / with an introduction by Janet Hurst.
pages cm
Includes index.
ISBN 978-0-7603-4565-8 (hardcover)
1. Goats. I. Hurst, Janet.
SF383.E98 2014
636.3›9--dc 3
2014001549

Printed in China
10 9 8 7 6 5 4 3 2 1

A MODERN MIRACLE,
THE:
Missouri Steam Washer,
NO HOUSE IS:
COMPLETE WITHOUT IT.
(over.)

BUMPER.
SUPERIOR
QUALITY:
HAND
MADE.

"THE FOX & THE GOAT" (See other side.)
KERR & Co's
ÆSOP'S FABLES.
EXTRA SIX CORD SPOOL COTTON

INTRODUCTION

Despite the bad press they get, goats are generally playful, affectionate animals.

Once the kids showed up, keeping goats became a way of life.

By Janet Hurst

As a young mother in the 1970s, I dreamed of a piece of land—a small farm where I could raise my son and share the wonders of rural living with him. Eventually we found such a place: a few acres, a little creek, a place to call our own. In the magazines of that time, especially *Mother Earth News* and *Countryside*, it seemed everyone had a Volkswagen, a backpack, and a goat. I decided country life could not possibly be complete without one. Or two. Everyone I knew tried to talk me out of it. My parents shook their heads and wondered where they had gone wrong. After all, I had been raised to be a "proper lady," not one wearing overalls and gum boots.

I found my prince one day at a flea market. I pulled money from my well-worn backpack and paid for the rights of ownership for my first goat—a billy, at that. I tucked him in the back of my Volkswagen bug. He was handsome, a young Nubian buck. I was in love. I named him Amos.

One goat is a lonely goat. Amos cried for companions and girlfriends. I complied with his wishes and purchased an Alpine nanny, in full milk. I named her Dolly Parton for two obvious reasons.

Now, there are lessons to be learned in the goat world. Lesson number one: a nanny sold in full milk is sold for a reason. Nobody is going to raise a goat, feed her all winter, and then sell her when she is in milk unless there is a pretty good reason. There was. Dolly Parton had horrible milk. This being my first goat, I didn't know any better and secretly wondered why everyone was so excited about this milk that tasted so terrible. One thing Dolly was good for was volume. She made lots of horrible milk, bless her soul. I decided to make cheese from it. Lesson number two: horrible milk makes horrible cheese.

I was not to be beaten, so I bought another goat. She gave the best milk I have ever had. It was rich, full of cream. She was another Nubian, the same breed as Amos. She had long, bassett-hound-looking ears, a Roman nose, and an udder that swayed seductively back and forth as she pranced through the pasture. No wonder Amos was smitten. His behavior changed the day this lovely creature cycled through her first heat. Amos turned from mild-mannered pest into a sight to behold. He made a little chuckling sound, began to urinate on his legs and beard, and courted the fair maiden in anything but a subtle manner. She was quite impressed by his antics, and nature took its course. Five months later, I found three babies in the barn, little Amos miniatures. They were perfect.

I began to make cheese from the milk of dairy goats. I spun fibers from Angora goats and learned all sorts of new skills, all centering upon goats. I was hooked.

Now, some thirty years later, I still find goats as captivating as in those early days. I learned what people have known for centuries: goats are practical animals, and, as I suspected, they do make a homestead complete.

I think these animals have wonderful personalities, for the most part. Goats are playful, happy creatures. A few minutes in the barnyard is good for the soul and a guaranteed stress reliever.

Goats do have long memories, however. If their feelings get hurt or they feel they have been wronged, they will adopt an attitude, and when the opportunity presents itself, they will settle the score. Males, particularly, are known for ornery barnyard behavior.

The primary factor in a happy existence with goats? Good fencing. It has often been said, "If you can throw water through a fence, a goat will get out of it!" A daily battle can begin if fencing is not properly managed. The animals will see a poor fence as a challenge, and the first thing you know they will be

Don't underestimate a goat's intelligence. Though they're typically agreeable, they may seek revenge if they're treated badly—and sometimes they're just mischievous.

Build a good, stout, goat-proof fence *before* the goats arrive.

feasting on rose bushes or chewing paint off the car. Avoid this by installing fencing before you bring the goats home.

These days, goats are finding their way back to the limelight. These animals are known as browsers (they eat brush, shrubs, and brambles), but they can also adapt to a small farm setting, consuming mostly hay and a ration of grain. They require shelter from wind and rain, and fresh water daily. Often a guard animal is run with goats to protect them from predators. Donkeys, llamas, and dogs are frequent pasture mates, keeping watch over the herd.

If there is one word that could be used to sum up a goat, it would have to be "utilitarian." They can adapt to any climate, from desert heat to frigid, high mountain ranges. Their ability to adapt to their circumstances, survive on less than desirable foodstuffs, and produce milk, meat, fiber, and other products makes them one of the most useful animals in domestication. There are more than 300 breeds of goats on record, from Arapawa Island goats to those native to the mountainous regions of France.

Records indicate goats were common on European sailing ships. Spanish explorers brought the first goats to the United States, and apparently some of these early animals escaped or were released into the areas now known as Texas and Oklahoma. These feral goats became known as "Spanish goats" or "brush goats."

As colonists began to inhabit the East Coast, Captain John Smith and Lord Delaware brought dairy goats to the newly forming country, and over the next few centuries, various goat breeds were imported from Europe. In the United States, though, goats have always taken a back seat to the cow. Small homesteads typically raised a few goats for their household milk supply, but dairy cows were the most common milk-producing animal.

In recent years, interest in artisan cheese has fueled the dairy goat industry, with cheesemaking plants and creameries dotting the landscape from Maine to California. Goat meat is also making its way into the culinary world, thanks to the ever-changing population of the United States. In addition, backyard farmers and rural homesteaders are seeing goats as a comfortable alternative to cattle. Because goats take up less room and eat less food, yet can provide milk, meat, and fiber, they are a better choice for a small farm.

In this country, it seems, our love affair with the goat is just beginning.

In the last few decades, artisan goat cheese has become very popular in the United States.

1 ★ HISTORY OF THE GOAT

Goat carriages at Coney Island in Brooklyn, New York, ca. 1900-1910.

Sketch of a figure representing the Year of the Goat, ca. 979 CE.

By Janet Hurst

In the United States, goats are experiencing a renaissance. It's an exciting time for goats and goat fanciers as this animal, which has been linked with humans since prehistoric times, has been rediscovered by the modern American.

Cave paintings, Biblical references, mythology, and folk tales depict goats at work and play. The chronicle of the goat begins in the Middle East—probably in what is now northern Iran—where the animal was first domesticated. Archaeological evidence traces goat farming back to nearly 10,500 years ago, when the animal was valued for its milk, meat, hair, hide, and dung.

The domesticated goat's scientific name is *Capra aegagrushircus*. Because the genus name is Capra, goat-related topics or qualities are sometimes called "caprine." The genus includes domesticated goats, wild goats, and several species of ibex (which are found in Europe, Asia, and Africa), one of which may be the ancestor of today's household goat, though it's unclear which one.

Apricornus ad occasum spectans & totus in zodiaco circulo deformatus: cauda & toto corpore medius diuiditur ab hyemali circulo: supposit? Aquarii manui sinistræ: occidit autem preceps: exoritur autem d rectus. Sed habet stellam in naso uná. Infra ceruicem unam. In petore duas. in priore pede uná. in priore eodem alteram in ersca pilio habet stellas septem. in uentre septem. in cauda duas. Et est omnino stellarum. xii.

Capricornus

Q uarius habet pedes i hyemali ciculo fixos: manum auté sinistram usq ad capricorni porrigés tergom dextrá iubæ pegasi pde cótingens: spectat ad exortus: qui cú ita sit figurat?: necesse est eú corpore prope resupinato uideri. Effusio aquæ puenit ad

Woodcut from Poetica astronomica, published in Venice in 1488. This work, which has been attributed to Hyginus (born in Spain in about 60 BC), was frequently copied and illustrated during medieval times. Hyginus' work recounts the legends of the constellations, the location of stars within them, and the motion of the Sun, Moon, and planets. This tradition was rooted in poetry and mythology; it contained minimal scientific content.

A Byzantine mosaic of a shepherd milking a goat, ca. sixth century CE. On display at the Great Palace Mosaic Museum in Istanbul, Turkey.

Segment of a wall decoration from the tomb of Amenhotep III, 18th Dynasty, 1390-1352 BCE Tomb 22 in the Valley of the Kings, Luxor, Egypt.

Seven Minoan terracotta animal figurines, including cattle, sheep, a dog, and a goat, ca. 2000-1700 BCE.

Relief depicting two billy goats eating from a grape vine, ca. sixth or seventh century: on display at the Pergamon Museum in Berlin, Germany.

Five carved horn heads from the headdress of a shaman of the Tlingit, from the Pacific Northwest coast, North America. They are made from the horns of mountain goats.

Goats appear in cave drawings, hieroglyphs, and petroglyphs. Cave drawings clearly indicating the presence of goats in Spain and France. Native Americans featured goats in their drawings, jewelry, and other adornments. Artists in the Near East also used images of goats in jewelry and sculptures and on pottery and ceramic seals.

Nomadic people traveled with small herds of goats. The animals were perfectly suited to rocky, desert terrain. Goats are herd oriented and will stay with their masters, traveling caravan style wherever the roads or paths may take them. What other animal could carry a pack, supply milk upon demand, and serve as a meat producer as well, all on a diet of scrub brush and thorns? Only the lowly goat.

The goat has withstood centuries of mockery, ridicule, and literal sacrifice, but the goat has also been honored and revered. A white goat with flowing hair is mentioned in the book of Solomon, no doubt referring to an Angora (the breed that produces mohair). "In the land of milk and honey"—a famous phrase from the Proverbs—is a reference to goat milk. In fact, references to the goat are made throughout the Bible. Goats are referred to as milk, meat, and fiber producers, and they are also the first chosen for sacrifice—implying they are the most valuable.

On the other hand, there's a parable in which the sheep are chosen to go on the right side of the Master, and the goats on the left. Reading this, it's hard not to think of the affable nature of sheep (they tend to do what they are told) and compare them to free-spirited goats, who have minds of their own.

Myths, Meat, and Milk

Why is it that people laugh when they think of this practical beast? Perhaps it comes from nursery rhymes and fairy tales depicting the goat as slightly

Major Russell Harrison, son of US President Benjamin Harrison, with his children and a goat cart in front of the White House, ca. early 1900s.

Villagers in Valais, Switzerland, tend to their goats, ca. 1890.

WORLD GOAT PRODUCTION

Selected Regions and Countries, 2008

Country/ Region	Total Animals (millions)	Goat Milk (MT)	Goat Meat (million MT)
Asia	511.3	8.89	3.4
Africa	294.5	3.2	1.1
China	149.37	0.26	1.83
India	125.7	4.0	0.48
Pakistan	60.00	N/A	N/A
Bangladesh	56.4	2.16	0.21
Nigeria	53.8	N/A	0.26
Sudan	43.1	1.47	0.19
Americas	37.3	0.54	0.15
Europe	17.86	2.59	0.012
Mexico	8.8	0.16	0.04
Afghanistan	6.38	0.11	0.04
Oceania	3.42	0.0004	0.018
USA	3.1	N/A	0.022
Saudi Arabia	2.2	0.076	0.024
France	1.2	0.58	0.007
UK	0.09	N/A	N/A
World	-----	15.2	4.8

deranged and sometimes demonic. Who doesn't remember the goats tromping across the troll's bridge? We need to remember, however, that these portrayals are fiction, and they fail to capture the true spirit of the animal. Goats are quite intelligent, and they are capable of planning ahead. Ask anyone who has been butted by a goat—they will certainly tell you it was premeditated.

Myths abound when it comes to goats. For the record, they do not eat tin cans. They will, however, lick the salt and sugar from the can and might consume a paper label. They are not dirty animals. They will not eat soiled hay or drink dirty water. A female goat does not have an odor. A male does have a strong musky presence when courting females, and he makes things worse by urinating on his forelegs and chin to attract females. Males and females both grow horns. A goat without horn growth is said to be naturally "polled." Males and females both have beards. This chin growth comes with age.

Most people do not realize that goat milk is the most widely consumed milk in the world and that goat meat is the most popular meat. Throughout the ages, goats have fed the masses. According to the Food and Agriculture Organization (FAO), the top producers of goat milk in 2008 were India (4 million metric tons), Bangladesh (2.16 million metric tons), and the Sudan (1.47 million metric tons).

When Americans speak of "milk," they assume the milk is from a cow. However, in other parts of

Goats on the street in Malaga, Spain, ca. 1938.

GOAT VOCABULARY

There is a language surrounding goat keeping and care. A list of terms:

Billy (slang) or **sire**: A male goat used for breeding purposes.

Buck: A male goat.

Chèvre: Fresh cheese, made from goat milk.

Colostrum: The first milk produced by a newly freshened doe. This is a thick yellow substance full of antibodies and is essential to the newborn kid's survival. This first milk is produced for about three days.

Doe: A female goat.

Fresh: A goat who is lactating is considered to be fresh. A goat who has recently had a kid and come into her milk is said to be freshened.

Kids: Young goats.

Nanny (slang) or **dam**: A goat mother.

Polling: Removing horns

Wether: A male goat that has been castrated.

Goats pulling a sled for Klondike-bound gold prospectors, ca. 1898. At the end of the journey, the goats were typically eaten.

GOAT FACTS

- All goats are ruminants, with a series of stomachs. They regurgitate their food, essentially chewing their cud (or reprocessing food that has already been eaten).
- Goat milk contains caprylic acid. This is what gives it the "tang" the milk is known for.
- A female goat must give birth in order to produce milk.
- Does cycle into estrus or heat every thirty days throughout the breeding season.
- A goat's pregnancy lasts for approximately 151 days.
- Twins and triplets are common in the dairy breeds.

the world, this is not necessarily the case. Figures show more than 72 percent of the world favors goat milk over cow milk. A favorite goat tale is that of a traveling milk vendor in Africa. This gentleman goes door-to-door, selling goat milk. The buyer brings out a pot, the goat is milked on the spot, and the currency exchanged. The herd is then driven to the next house and the routine begins again. Milk does not get any fresher than that!

Goats have long been known as the cheesemaker's animal. Many farmstead operations still produce cheese directly on the farm. Small creameries produce cheese unique to each farm, region, and breed, allowing for the development of unique and signature products.

Milking, in particular, is an intimate relationship. Farmers' lives intertwine with their animals. There are favorite goats, which present themselves for a scratch or an extra bit of grain. This is another reason goats have withstood the test of time—they are affectionate beings, and they enjoy human company. Once a human is truly bitten by the goat "bug," there is little chance of recovery. The only cure? A goat or two.

Goat milk is often said to be naturally homogenized. Actually, the fat globules stay suspended in the milk, which is slightly more viscous than cow's milk. In cow's milk, the cream all rises to the top, forming a thick layer. With goat's milk, the cream stays integrated within the milk. While some cream rises, it is more difficult to skim, making buttermaking from goat's milk a more difficult task. A cream separator (which is a centrifuge) is most often used to separate the cream from the milk. Due to the nature of the milk and the distribution of the fat, goat milk is often said to be easier to digest than cow's milk. Many stories exist regarding sickly babies becoming healthy and robust after being fed goat milk. Indeed, goat milk has been suggested as a cure for everything from stomach ulcers to poison ivy. (See Chapter 9 for more on this.)

Depending upon their locale and intended purpose, various goat breeds have been developed to fill specific needs. Certain breeds of goats are kept mainly for milk and cheese, and others are kept for meat or fiber. (See Chapter 2 for an introduction to goat breeds.) In addition to these products, other parts of the goat can be used in some manner. Hides can be tanned and stretched to create the surface for drums, kid skin for gloves, hooves for gelatin, stomachs for waterproof vessels—the list goes on and on. Is it any wonder goats were domesticated early on?

Caption reads "Goats milked while you wait—in a crowded market place in old Palermo, Sicily." 1906.

2 ★ GOAT BREEDS

A herd of Kashmir goats in Ladakh, India.

By Janet Hurst

There are more than 300 breeds of goats on record, and most are categorized by their primary use: fiber, meat, or milk. Fiber and milk goats are dual purpose in that they can be consumed for meat or used for their better-known qualities. According to the 2013 Sheep and Goat Report from the United States Department of Agriculture National Agricultural Statistics Service (USDA- NASS), the United States is home to more than 360,000 dairy goats, 136,000 Angora goats, and 2,315,000 meat (or other) goats, for a total of 2,811,000 goats in the United States alone.

Angora kids

Fiber Goats

As fiber animals, goats are unsurpassed for their luxurious mohair and cashmere production. Mohair comes from Angora goats and Cashmere (or Kashmir) goats produce cashmere. The garments of royals were often woven of fabric spun from these coveted yarns.

Angora goats are sheared twice a year, in spring and fall. The fleece is gathered, combed, and washed to reveal glistening lengths of curly locks. Mohair must be handled with great care. First the fleece is skirted to remove vegetation and debris. Then it is carefully washed. If extremely hot water or agitation is used while washing, the fibers will come together, forming a thick, felt mat. Therefore, soaking in a mild detergent and simply spinning the locks out in a washing machine is the best method of cleaning. The mohair is then placed on screens to dry. The typical Angora goat has a snowy white coat. More recent genetic selection has focused on breeding for natural color with black-, silver-, and red-coated goats prized for their unusual shades of fleece. Mohair is known for its durability and is used in fashion, home décor, and industrial manufacture.

Cashmere originated in Kashmir, India, hence the name of the goats that produce cashmere fiber. Fibers should be gathered by a labor-intensive combing process, though overall shearing of the animal is common as well. The combed hair is of a finer quality, resulting in a superior fiber. The process of hair collection is done in the spring, when the animals naturally lose their winter coats. The fine wool or Cashmere fiber is spun into yarn to produce fine garments, highly prized for their durability and warmth.

A Cashgora is a cross between a Cashmere doe and an Angora buck. Appearance is similar to the Angora, with curling locks and shimmering coat. This type of fiber has fewer secondary (guard) hairs than cashmere, which makes for a very soft, down-like yarn.

Dairy Goats

These animals are known for their high level of milk production. All goat breeds produce milk, but these breeds are raised primarily for milk. Goats are typically milked twice a day, and milk is measured in pounds. One gallon of goat milk weighs a little more than eight pounds. One gallon of milk will yield approximately one pound of cheese. Females in milk are said to be "fresh." The typical lactation cycle is nine to ten months. This highest level of milk production is between two and three months into the freshening period.

The Saanen is typically an all-white or sometimes sable-colored goat. A large-framed animal, the Saanen ranges in weight from 120 to 150 pounds. Named for the Saanen valley in Switzerland, they are the most prolific of milk producers, often giving more than 1½ gallons per day. These goats are known for docile behavior, which makes them a favorite breed for dairy herds.

(Anglo) Nubians are known for their Roman noses and long, floppy (beagle-like) ears. A full-blooded Nubian can have an ear length of twelve inches or more. A large-framed animal, slightly smaller than the Saanen, a doe weighs in at about 135 pounds. This breed was developed in Great Britain by crossing British milk goats with animals from Africa and India. The Nubian is not the highest in the volume of milk production, but they are known for the highest fat content. Cheesemakers value this trait, as high butterfat levels mean more solids in the milk. More solids translate to higher volumes of cheese per pound of milk. Nubians are quite vocal. Some goatkeepers prefer a quieter breed. However, their personality and noble presence makes them a favorite homestead addition. Nubians range in color from pale tans to dark blacks with brown or white markings. Some have spots, and that trait is prized among goat breeders.

Alpines are a bit small than Nubians or Saanens. Some say milk from this goat has more tang to it than other breeds. For those who are making cheese or drinking milk, this can either be a desirable or undesirable trait, depending on personal preference. Originally from the French Alps, Alpines are a midsize breed with does weighing approximately 125 pounds. They are prolific milk producers, competing with Saanens in overall milk production. Colors are specified by the following terms: Cou Blanc, translates to white neck, black hindquarters; Cou Clair, clear neck, black hindquarters; Cou Noir, black neck; Chamoisee/Chamois, brown body with black face, feet, and legs.

The Oberhasli originated in Switzerland. This is a midsize goat with distinct markings. The name Oberhasli translates to "Highlander." The Oberhasli is typically a bay

Saanen goats

A Nubian goat

Oberhasli goats

A LaMancha goat

An Alpine goat

A Toggenburg goat

(reddish-brown) color with a black dorsal stripe down the backbone and black tipping on face and ears. It's not as common as some of the other dairy breeds, though it's a good-natured animal and a good milk producer, offering milk with a slightly sweet taste. The Oberhasli is listed as a recovering breed with the Livestock Conservancy.

LaManchas are sometimes said to have no ears. This is not so. They have tiny ears lying close to their heads—sometimes these are known as gopher ears. This breed has a wonderful personality, and they are second in butterfat production, only outranked by Nubians. The LaMancha is an American breed and the only recognized dairy goat breed to originate in the United States. Does weigh about 30 pounds.

Originating in Obertoggenburg, Switzerland, the Toggenburg is of medium build, a good milk producer, and known as being the oldest breed of dairy goats on record. Colors range from fawn to dark brown with white markings on the ears and face.

Nigerian Dwarf goats at the Norman J Levy Park and Preserve, Merrick, New York, 2012

Dwarf Goats

The African Pygmy is the smallest of the goat breeds, weighing in at 50 to 75 pounds at maturity. This breed originated in West Africa and was imported to the United States in the 1950s. First shown and kept as zoo animals, this miniature soon caught the imagination of the pet industry. Though these goats were first introduced as an anomaly, their small stature allowed their introduction into urban and small farm settings. Despite their small size, these animals can produce a surprising amount of milk.

The Kinder is a midsize goat, weighing in at 100 to 125 pounds. This breed is relatively new, with records dating back to only 1985. Kinders are unusual in that they are a cross between a Nubian (fullsize) female with a Pygmy male. The end result is a short, stocky animal with a wonderful disposition. These goats inherited the Nubian traits of high-quality milk with a moderate rate of production. However, they are thrifty feed consumers due to the Pygmy genetics, and thus a good choice for the small land holder.

▲ Kinder goats

◀◀ An African Pygmy goat

The Nigerian Dwarf weighs in at 75 pounds. These pint-size mammals are known for backyard milk production. First introduced in the 1980s, this breed was developed in West Africa. Nigerians are a dual-purpose breed, providing both milk and meat. Nigerians are considered rare and are listed with the Livestock Conservancy.

Meat Goats

Like their bovine counterparts, there are specific breeds that are best for meat production. Among those are the New Zealand Kiko, the Spanish Meat goat, the South African Boer goat, and the Tennessee Meat or Fainting goat. The meat goat breeds are known for their rapid rate of gain, the quality of their meat, and their ability to convert feed into body mass.

The New Zealand Kiko hails from New Zealand, where they were developed by crossing wild goats with tame dairy goats. Developed in the 1980s, the breed is relatively new. However, due to inherent traits, such as low maintenance, parasite resistance, maternal instincts, and overall rate of growth, these animals have become prevalent in meat goat production.

The Spanish Meat goat came to the United States via Mexico. These goats are a mixture of many different types of goats with influences of various breeds apparent. It appears they have captured the best traits of their ancestors. Over time, their hardy constitutions, rate of gain, and overall ease of maintenance have made them a favorite among

A Tennessee Meat goat, also known as a Fainting goat

Close-up of a Boer goat

A herd of Boers

meat goat producers. This animal is on the Livestock Conservancy watch list.

The South African Boer is one of the most commonly recognized meat goats. Known for their white bodies and tan, black, or red head markings, they are found throughout the world. Boers have long, basset-like ears, similar to the Nubian. Weighing in at 90 to 130 pounds, these animals are bred for rate of gain and overall quality of meat cuts.

The Tennessee Meat or Fainting goats are known as Mytonic goats, Tennessee Meat goats, Wooden Leg goats, or Fainting goats. These animals possess an unusual trait. When frightened or startled, their muscles freeze and they fall over in what appears to be a faint. Actually, the animal's muscles are simply contracting, and it does not lose consciousness. This hereditary condition is known as mytonia congentia. Along with this peculiarity, this breed is known for its overall hardiness and rate of gain. The breed is listed as threatened by the Livestock Conservancy. As a side note, one may ask what it takes to induce "fainting spells" in these animals. Reportedly, a loud noise or backfire from a car is enough to take this goat down. However, a friend reported she went outside in a rather loud, bright yellow sun hat. Her goats took one look and fell right over.

New Zealand Kiko goats

Arapawa Island goat

Heritage Goats

There are several heritage goat breeds listed with the Livestock Conservancy. The Oberhasli, Spanish, and Mytonic goats have been previously noted. Other breeds that are in decline include the Arapawa Island goat, a feral goat that is native to New Zealand and directly descended from animals brought by Captain Cook and other settlers in the eighteenth century. These wild goats are the last known representatives of the Old English breed.

The San Clement is also on the Livestock Conservancy's critical list due to a declining population—at last count there were 300 animals left. This breed is feral and ran wild on the California island of San Clement until the 1980s.

The Golden Guernsey, a goat once popular with families in the United Kingdom, is also on the watch list. Animals on this list are being watched but are not in danger of extinction at this time.

A Golden Guernsey

3 ★ GOATS IN MYTHOLOGY

An astronomy chart from Samuel Leigh's ca. 1825 collection *Urania's Mirror* (plate 25) shows the mythological figure of Capricornus, a half-goat, half-fish creature with an ancient Near Eastern lineage. The figure is said to be derived from the image of the Sumerian god Enki, the deity of creation, water, beer, and mischief. Cylinder seals from about 2100 BCE and Babylonian star catalogs from about 1000 BCE record the Capricorn, or Goat-Fish, creature. In Greek mythology, this constellation is sometimes identified as Amalthea, the goat that fed baby Zeus after his mother, Rhea, rescued him from a hungry Cronos. Unsurprisingly, Capricornus is also linked with Pan, who once escaped from Typhon by growing a fish's tail and swimming away.

This statue of Pan, the Greek god of fertility, is in the Musei Capitolini in Rome, Italy.

By Darwin Holmstrom

Joseph Campbell, the renowned scholar and author who wrote *The Masks of God*, *The Hero with a Thousand Faces*, and other works on the intersection of anthropology and myth, spent his lifetime trying to help people understand that mythology—which he defined as "other people's religion"—could have as much impact on the physical world as concrete reality itself. Campbell believed, in fact, that to understand mythology was to understand the very essence of human nature.

And to really understand mythology, you have to understand the role of one of the primary mythological symbols: the goat.

From Sacred Goat to Goat-Man

The goat holds primary significance in human mythology because this charming critter holds primary significance in human culture itself. It was the goat that made possible the very notion of civilization. About 11,000 years ago, the goat became the first animal to be domesticated as livestock; it was our ancestors' first reliable source of protein. It was their first source of meat and their first source of dairy. The goat's domestication saved our ancestors from the brutal existence of the hunter-gatherer and enabled the formation of human community.

The Sumerians, one of the first tribal groups that developed what could be called "civilizations," acknowledged the importance of the goat by making it a holy entity, an object of worship. As civilizations advanced, so too did the veneration of the goat as a mythological symbol. By the time ancient Greece attained the status of a complex civilization, around the time of the poet Homer, in 800 BCE, the goat was entrenched as a mythic symbol. Greek theatre developed as part of a festival called "City Dionysia," an orgiastic extravaganza celebrating Dionysus, god of wine, winemaking, and

A painting of Pan, ca. 1592, in the collection of the National Gallery of Victoria, Melbourne, Australia

ecstasy. Some of the first dramatic plays in human history were presented at City Dionysia festivals. The primary form of dramatic plays during this period was the tragedy, and the Greek word for tragedy—*tragoidia*—evolved from the words *tragos* (he goat) and *aeidein* (to sing). In other words, "tragedy" literally translates as "goat song."

These dramas started as simple affairs, little more than ritualistic sacrifices of goats to Dionysus. (The goat was considered sacred to Dionysus and was the most valuable thing a human had to offer the god at that time.) These dramatic presentations grew into elaborate affairs that eventually included the obligatory Greek chorus, dressed in goat skins, of course. The presentation of these plays ultimately became contests in which the winners were awarded—what else?—goats.

From these celebrations, goat veneration evolved into an anthropomorphizing of the goat as god, in the form of Pan. Pan, whose earliest appearance is in Pindar's Third Pythian Ode, was sometimes said to be the son of Zeus, though he is more commonly thought to be the son of either Hermes or Dionysus. It's tricky to find a first-person source to say for certain, so we'll go with Dionysus, since it fits our narrative. Pan's parentage might be open for debate, but his personality and physical appearance are generally agreed upon: he's depicted as a man-goat—goat from the waist down, human from the waist up—with a pair of goat-like horns adorning his shaggy head, and a massive mutton member. He was, after all, the god of fornication and fertility. The sexual exploits of Pan and his prodigious penis were the stuff of legend. He most famously seduced Selene, the moon goddess, but he wasn't exactly discriminating.

The goat has an honored place in other religions and cultures, as well—for instance, in Norse mythology, a goat named Heidrun eats from Yggdrasil, the tree that supports the world, and produces mead (instead of goat milk) every day for the warrior spirits at Valhalla. Thor's chariot is pulled by two goats, whose names mean "Teeth-snarler" and "Teeth-grinder."

On the November 8, 1919 cover of *Pan*, a literary magazine, a party girl rides the Greek god.

This Liebig card is illustrated with a mythological scene of fauns and a goat dancing, 1896.

The Goat of God

The Sumerians provide us with our earliest example of goat deification, but their descendants, the Semitic tribes of the Near East, developed monotheistic religions that squeezed out the goat as god, ultimately turning the useful little beast into a demonic symbol. Still, the goat had a good run in the Judeo-Christian culture.

Since the goat was the primary food source for early Semitic tribes, it's no surprise that their rules and regulations encouraged its use as a dietary staple. Most likely the Levitical laws regarding what could and could not be eaten by early Hebrews were based on observing what early Hebrews did and did not eat. The goat be cloven of hoof and cheweth the cud; therefore being cloven of hoof and cud chewething be good things. The camel and the ass cheweth the cud, but they be not cloven of the hoof; therefore not being cloven of the hoof most likely be a bad thing. The swine be cloven of hoof, but he cheweth not the cud; therefore the swine be really unclean, and he be righteth out of the running when it cometh to his role as a dietary supplement.

In this image of Adam and Eve, taken from the Holkham Bible Picture Book, published in England, ca 1320-1330, Eve is spinning and Adam digs the earth.

When it came to sacrificing an animal to God, the goat was the bomb. A goat was often a Hebrew's most important possession, and thus the most meaningful symbolically. Levitical law actually allowed for any one of three defect-free, kosher animals to be used as a sacrifice: a bullock, a goat, or a lamb. A goat was an obvious choice because it was the most accessible—pretty much everyone owned at least a goat or two.

The word "ram" was used rather freely in the Old Testament, and could refer to either a male sheep

Thor's Fight with the Giants, is an 1872 oil painting by Swedish artist Mårten Eskil Winge (1825-1896).

In Norse mythology, two goats pull Thor's chariot. Every night he eats them for dinner, and the next day he resurrects them with his hammer.

or goat. This ambiguity is seen throughout the Old Testament and is especially pronounced in the prophetic visions of Daniel. Given the importance of the goat for the ancient Semitic tribes among whom these mythic stories originated, the goat is the more likely source of Godly protein.

With the advent of Christianity, the goat again played an important role, especially when the early Christians began writing down the Christ story in order to help spread Christianity to the Greeks. As we've seen, the Greeks were goat-crazy, and in the original texts, Christ is presented as the Goat of God rather than the Lamb of God. To the Greeks, the Lamb of God would have represented personal redemption, but the Goat of God would have held deeper symbolic meaning, representing personal cleansing and payment for one's sins in addition to personal redemption.

This mosaic illustrates the parable of Jesus (the Good Shepherd) separating the sheep (on his right) from the goats (on his left); from the Byzantine School, sixth century.

Somewhere between the first-century marketing of Christianity to the Greeks and the printing of the Great Bible (the first authorized, English-language translation of the Bible, published in France in 1539), the Goat of God had become, irrevocably, the Lamb of God. By the time the first Great Bible came off the printing press, Calvinists had "reformed" Christianity into something so sexually repressed that it held no place for a horny little critter like the goat. The goat, it seemed, had been 'scaped.

Sheep Go to Heaven, Goats Go to Hell

Actually, the precocious caprine wasn't exactly 'scaped, at least according to the original meaning of the word. Originally the 'scaped goat referred to the escaped goat, a term that originated with the Hebrew ceremony celebrating the Day of Atonement. The Hebrews selected two goats for this ceremony, one unlucky bugger to be sacrificed and one lucky bugger who would be taken out to the wilderness and released: hence, the 'scaped (or escaped) goat, referred to in Hebrew as the Azazel goat. On the Day of Atonement, the killed goat was given to God and the Azazel goat's task was to take the sin of the people away. Eventually this became taking the sin to Satan in the Christian tradition, and really marked the beginning of the end for the unsuspecting goat's heroic status in mythology.

This illustration depicts a Freemason procession honoring Baphomet, ca. 1890.

The blame for this lies squarely at the feet of Saint Paul himself. Yes, Paul did indeed write some beautiful poetry, but, well, there's no delicate way to put this—Paul really, really hated women, and, in fact, hated sexuality in general. The Episcopal priest and theologian John S. Spong has theorized that Saint Paul was in fact a "self-loathing and repressed gay male." For this, his Very Reverend Spong has been threatened with excommunication from his church and worse, but he makes some good points. Unfortunately for women—as well as our hapless goat—Saint Paul was also the single most influential person in the history of the Christian tradition. His views helped form a religion that canonized the attitude that sex is icky—an attitude that spread across most of the Western world. When half the planet thinks that sex is icky, there's not much room for the horny goat as anything but a mythological symbol of pure evil. And that's exactly what the goat became in Western culture.

By the time of the Inquisition, the goat had become the symbol for Satan. When the Catholic Church decided to wipe out the Knights Templar, priests simply accused the Knights of worshipping Baphomet, the Sabbatic Goat that represented Satan. The name "Baphomet" is actually a slur against Muslims dating back to the First Crusade (1096-1099). Raymond of Aquilers, chronicler of the First Crusade, originally called mosques "Bafumarias." Later the name "Bafometz" was the name ascribed to the Islamic prophet Muhammad. This eventually morphed into the mythic image of Baphomet, a demonic creature with cloven hooves, female human breasts, and a goat's head. This symbol became so associated with Satan that it was adopted by Anton Szandor LaVey when he started the Church of Satan in San Francisco in 1966.

Witches' Sabbath is a 1789 oil painting by Spanish artist Francisco José de Goya y Lucientes (1746-1828). Goya painted several scenes of witchcraft, which are believed to be a satirical comment on religious and political oppression.

In a predominantly Christian society, there really is no coming back from something like that. One would have thought that it was all over for the poor goat, at least mythologically speaking, but a funny thing happened between the founding of the Church of Satan and today: Western society became much less Christian. Today there are far more Muslims and members of other religious traditions in North America and Europe than there were in 1966. The rise of other religions in Western culture has led to a corresponding increase in the market for halal and kosher goat meat, and the goat has once again become an important source of protein. Once the goat has resumed a respected place in the food chain, it will only be a matter of time before it resumes its rightful place in the mythological pantheon.

4 ★ CAPRINE CULTURE

This photograph shows Pablo Picasso (1881-1973), leaning on a podium in front of his goat sculpture *La Chevre*, inside the Salon De Mai, Paris, France. 1952.

Guido da Siena's *Majesty*, a gothic fresco ca. 1230, is found in the Hall of Pillars in the Palazzo Publico in Siena, Italy.

The gilded silver handle of this vase depicts a winged ibex; Persian, from the Achaemenid period, 6th-4th century BC.

Goats in Art

From the earliest of rudimentary carvings to crafts and sculptures to modernist art, the goat appears: sometimes as a symbol of an agricultural existence, sometimes as a nurturing force, and—depending on the religious era—as a symbol of good or evil.

In ancient Greek mythology, the goddess Amalthea, who nurtured and fed the infant Zeus, is often depicted as a goat. As time passed, and the church became the primary patron of the arts, depictions of goats took on a decidedly different religious symbolism.

However, as the arts flourished during the Baroque period, dramatic, classical themes were frequently revisited through stylized throwbacks to the ancient Greeks.

During the Industrial Revolution, as people shifted into cities and factories, artists painted idyllic scenes of rural life, replete with goats.

Before long, the processes of industry gave rise to mechanically derived pictures, and artists took note. Photography helped us understand the mechanics of motion—and who better to study than our longtime friend the goat?

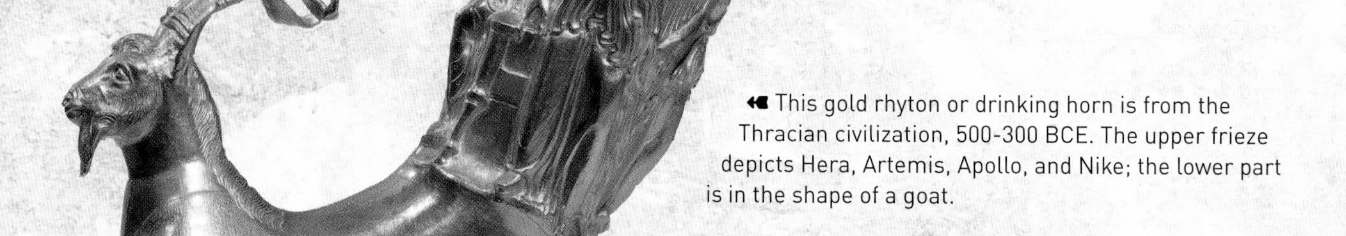

This gold rhyton or drinking horn is from the Thracian civilization, 500-300 BCE. The upper frieze depicts Hera, Artemis, Apollo, and Nike; the lower part is in the shape of a goat.

This goat head is part of the bronze sculpture known as the Chimera of Arezzo, which dates to the Etruscan civilization, 5th-4th century BC. A main part of the statue is a lion's body with a (reconstructed) serpent for a tail.

Countryside (ca. 1870), oil on panel, by Ernesto Bertea, hangs at the Galleria Civica D'Arte Moderna E Contemporanea.

This vase, called an "oinochoe," was used for serving wine. It was made in the Greek cities of southern Asia Minor, ca. 625-600 BC. The horizontal rows of decoration filled with wild goats and floral motifs reflect influence from the civilizations of the ancient Near East.

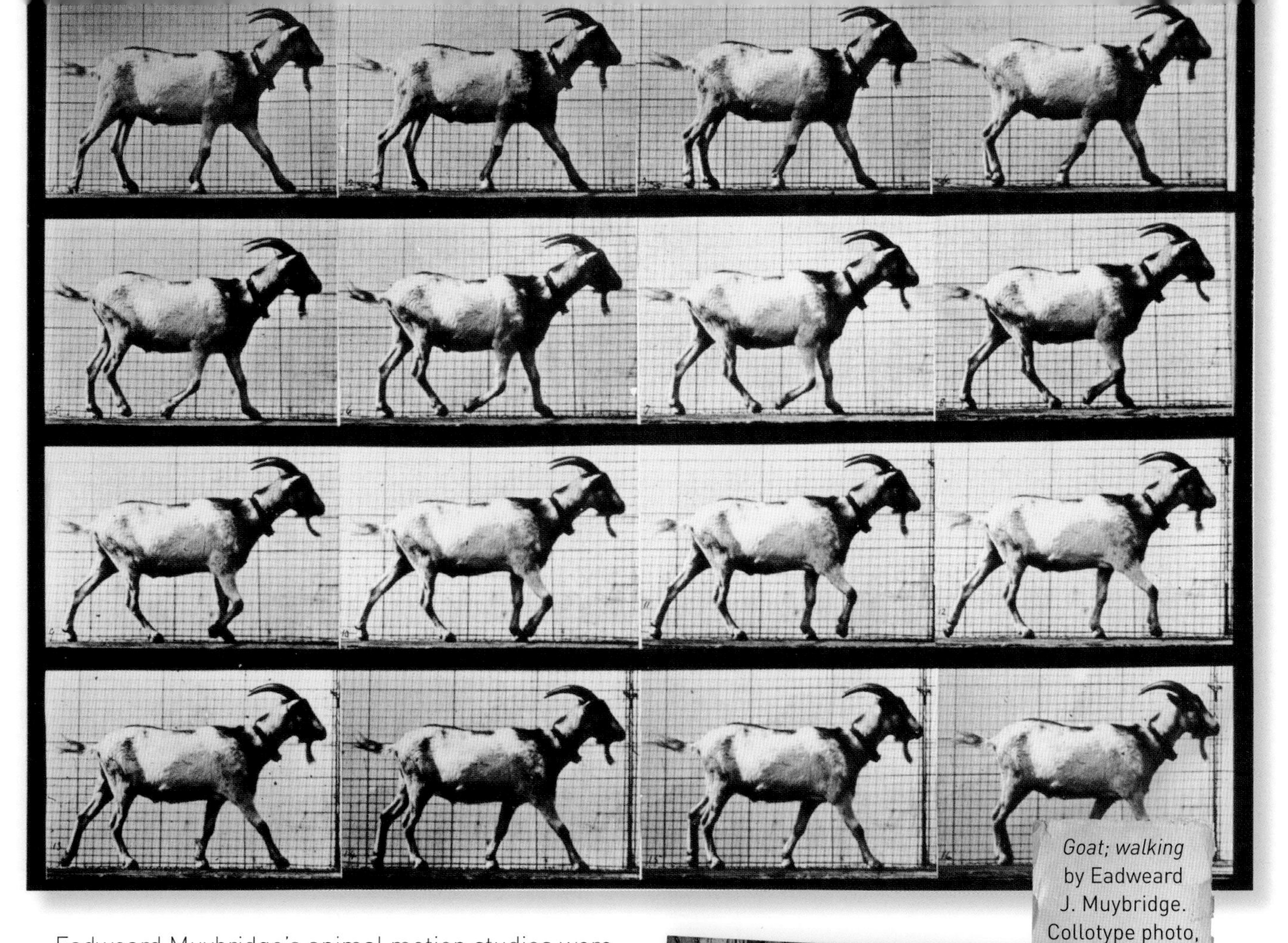

Goat; walking by Eadweard J. Muybridge. Collotype photo, ca 1884

Eadweard Muybridge's animal motion studies were precursors to motion picture film. Through his studies he was able to break motion down into individual steps of movement that were not distinguishable by the naked eye. His study of the horse's gait was the first to document that a horse does indeed leave the ground entirely during a gallop.

These new methods of viewing everyday life led artists to experiment with the representation of figures in motion and from multiple points of view. This goat's leisurely stroll (and other similar work by Muybridge) directly influenced Marcel Duchamp and his famed *Nude Descending a Staircase, No. 2.* This work is described as both Futurist and Cubist: Cubist because the subject is depicted from multiple points of view and it is a contextual observation of the process of descending the stairs, and Futurist because it portrays the world in constant movement.

Put another way—were goats the real driving force behind Cubism?

Nude Descending a Staircase, oil painting by Marcel Duchamp, 1912

Monogram, mixed media by Robert Rauschenberg, 1955-1959

Goat, by Manjit Bawa, 2003-2004

Goat sculpture on lawn, date and artist unknown

In Robert Rauschenberg's 1959 artwork *Monogram*, found objects are pulled together in a collage. Known for his "combines," Rauschenberg draws upon the symbolic history and also the playfulness of the goat in commenting upon the status and overly serious nature of artmaking at the time, particularly that of Abstract Expressionism.

The goat continues to appear in art throughout the twentieth and into the twenty-first century, and its presence is almost universally a reference to mythological, religious, and social symbolism. Our world changes, and art forms and techniques evolve with our shifting cultural landscapes, but our relationship with goats remains, and the archetypal image of the goat is forever etched in our visual lexicon.

Goat dancing at a Christmas party held by the Children's Zoo hostesses of the London Zoo, Regents Park, 1954

Goats and Music

By Dennis Pernu

It's probably a stretch to call goats the audiophiles of the barnyard, but the fact is, our caprine friends have a number of notable associations with music. To modern *rawk*-oriented minds, the most obvious of these connections is as a frequent totem of heavy metal, where goats have for decades been featured on album covers, in song and LP titles, and even in band names. But the goat's role in music runs much deeper—all the way to ancient times.

Though the goat has been associated with music since ancient times, goats themselves have a somewhat harsh song.

In Greek mythology we find the goat-like deity Pan, whose name is the Greek root for "all." Pan was adept at pastoral music, a rustic sort of storytelling set to melody. In other words, Pan was kind of like Tom T. Hall with goat horns and hindquarters. Like many heavy metal dudes, Pan was also a rather randy sort (as befit his apparent genetic makeup). In fact, his instrument of choice was the *pan flute* (natch), a wind instrument composed of seven reeds, one of which was formerly a nymph whom Pan had pursued with no success, thus, no doubt, becoming a source of shame among countless generations of billies to follow. In expanded form, the pan flute would, of course, much later be popularized by a Romanian fellow named Zamfir whose late-night television commercials were seemingly ubiquitous in the 1970s and 1980s. In a moment of goat-like hubris, Pan even challenged Apollo, master of the stringed lyre, to a sort of Olympian battle of the bands. Pan's new pal, Midas, still bummed about having turned his daughter into a pillar of gold, and having thus renounced wealth, was the only one present to cast his vote for the flutist; for

In the classic children's book *Wind in the Willows* (1908), by Kenneth Grahame, Pan appears as "The Piper at the Gates of Dawn."

An album by Gheorghe Zamfir, modern master of the pan flute

his temerity, Midas was awarded the ears of an ass. That guy couldn't catch a break.

At the end of the day, Pan didn't fare much better. Not only is he cited by the historian Plutarch as the only Greek god to die, but his demise became allegorical among medieval Christians trying to square Greek and Roman mythology with certain Christian notions. It was at this point that the fertile goat began to displace winged reptilian creatures as representations of Satan, demons, and the like among Christians, and not surprisingly find his way into the good graces of pagan and Satanic rituals. Much later, Pan was co-opted by nineteenth-century Romantics such as John Keats and by twentieth-century neopagans. (An interesting side note: while neither Romantic nor neopagan, Kenneth Grahame, in his 1908 children's lit classic, *The Wind in the Willows*, featured Pan as "The Piper at the Gates of Dawn," a title that Pink Floyd would appropriate for their 1967 debut album.)

Pan—and by extension, goats—breached other aspects of popular culture: note the appearance of goats in nineteenth-century opera. German composer Giacomo Meyerbeer's *Dinorah* caused uproar in New York City when performances featured a live goat. Italian Francesco Cilea penned an aria about a goat, and one form of vibrato noted during this period is still known as *chevrotement*—literally "goat's trill." In her book, *Sing! Vocal Technique, Vocal Style* (2006), Elisabeth Howard cautions that even today "many treatises on singing issue stern warning" against chevrotement.

Neopagans and their love for our libidinous, Greek, demi-goat friend would influence Satanists to a degree, and ultimately we can thank the Satanists for the goat's starring role in heavy metal.

Religious groups from Bahá'i to Wiccans to Taoists to Mormons have incorporated the pentagram in their faith in one manner or another. Satanists followed suit, but with a bit of caprine panache, nestling a fearsome-looking goat's head into an inverted pentagram. It's easy to imagine the use of a goat was inspired at least in part by Christ's parable of the sheep and the goats (Matthew 25:31-46), in which the son of God, on Judgment Day, places the abiding sheep at his right hand and casts the recalcitrant goats into eternal damnation, intoning, "Depart from me, you cursed, into the eternal fire." (Sounds delicious with a bit of rosemary, garlic, lemon, and olive oil.) Given the fact that the Church of Satan later trademarked the symbol as the Sigil of Baphomet in 1983, it seems more likely their goat use was influenced by occultist Eliphas Lévi's nineteenth-century drawing of a bipedal creature that he called the Sabbatic Goat, but which later became referred to as Baphomet, for an idol allegedly worshipped by the crusading Knights Templar centuries earlier. For his part, Lévi claimed to take inspiration from the Witches' Sabbath (immortalized by Spanish painter Francisco Goya) and from the Egyptian Goat of Mendes. Scholars agree Baphomet was a bastardization of the French name for Muhammad, and that some Knights Templar had brought Muslim ideas home from the Crusades, incorporating them into their Christian beliefs. Allegations of such heresy were all that King Philip IV needed to (wait for it) scapegoat the knights and thus renege on debts the crown owed these mercenaries.

Whatever the case, a goat head just plain works well in an inverted pentagram, with enough points for two horns, two ears, and one chinny-chin-chin—a fact not lost on the New Wave of British Metal band Venom, who lifted the Sigil of Baphomet almost wholesale for the sleeve of their 1981 LP, *Welcome to Hell*. In 1993, D.C.-based Pentagram reissued their debut album with cover art nearly identical to Venom's (apparently flying under the radar of the Church of Satan's legal staff). Incidentally, Venom apparently knew when they had a good thing going—a goat also graced the cover of their 1982 sophomore effort, *Black Metal*.

Bands and fans of certain heavy metal subgenres—e.g., black metal and death metal—are often practicing Satanists, but more often than not, Satanic signifiers in heavy metal are, at worst, high camp—conceits intended to rile more puritanical elements of

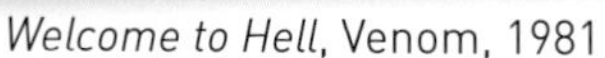

Welcome to Hell, Venom, 1981

Witchcraft Destroys Minds & Reaps Souls, Coven, 1969

Bathory, Bathory, 1984

society (and thus sell more records, downloads, and concert tickets). Of heavy metal's practicing Satanists, none are more steadfast in their beliefs than those from Scandinavian nations, many of whom resent that Christian churches long ago displaced their Nordic cultures' own mythos, in which goats play significant roles (pulling the chariot of Thor, for example, and lactating hot mead for Viking warriors living it up in Valhalla). Finland gives us not only the band Archgoat, but also Impaled Nazarene, who devote one track on each album to the goat (e.g., "Goatzied," "Goat War," "Goat Seeds of Doom," "Oath of the Goat").

But perhaps no Scandinavian metal act is more associated with goat lore than Swedish band Bathory, named for the infamous Hungarian countess said to have bathed in the blood of virgins. True to the DIY spirit, frontman Quorthon set about to design the sleeve of the band's 1984 self-titled debut himself. With the requisite goat image chosen, and using rub-on letters of the pre-desktop publishing age, Quorthon found himself one "c" short for the track listing on the back sleeve, thus "Necromancy" was henceforth known as "Necromansy." Quorthon had also hoped to print the sleeve in black and gold inks. Upon learning that gold ink was cost prohibitive, however, he allowed the printer to choose a tone to approximate his original vision. The resulting garish yellow and black cover was replaced with black and white art after the initial run of 1,000 records. Today, the *gulageten* (yellow goat) LP is a collector's item.

The Scandinavian black metal scene's proclivity for goats aside, nowhere in heavy metal—nay, in music—is the goat more prevalent than in a rite known as "throwing the goat," the act in which the pinkie and index fingers are raised skyward and the ring and middle fingers folded down, covered (this is important) by the thumb, theoretically symbolizing the head and horns of a goat.

Exactly who started this tradition has been the subject of debate for decades. Naturally, Gene Simmons of KISS has tried to stake claim, but the late metal vocalist Ronnie James Dio, who fronted Rainbow before replacing Ozzy Osbourne in Black Sabbath, is most often credited, an attribution which he modestly shunned before his death in 2010, explaining that he learned the hand sign from his Italian grandmother, who used it to ward off the evil eye, which had the power to wilt a man's virility (perhaps suggesting some sort of goat horn derivation).

It would appear that Dio's modesty was far from false—and that throwing the goat is, in fact, not even of heavy metal derivation. The first photographic appearance of the now-ubiquitous gesture in rock music appears on the back cover of *Witchcraft Destroys Minds & Reaps Souls*, the 1969 debut by Chicago psych-rock outfit Coven. LP and band name aside, Coven foretold many future trappings of heavy metal, with song titles such as "Satanic Mass" and "Pact with Lucifer." In a strange twist, the LP also featured a song called "Black Sabbath"

The Beach Boys' *Pet Sounds* (1966) showed the band feeding goats at the San Diego Zoo.

In both the cover and the picture sleeve of the Beatles' *Yellow Submarine* (1969), John Lennon appears to be throwing the goat—but he's not doing it right.

and a bassist named Oz Osbourne—mere months before a Manchester, England, band called Black Sabbath fronted by a nutter vocalist named Ozzy Osbourne burst onto the scene, perhaps explaining why the Mancunians are often credited with concocting the gesture. Interestingly, on a parallel planet (almost literally), George Clinton and Bootsy Collins of Parliament Funkadelic were trading the hand signal as the "P-Funk sign," nearly oblivious to contemporary heavy metal happenings.

It's worth noting that months before Coven released their debut, the Beatles issued *Yellow Submarine*, whose Heinz Edelmann–drawn cover art featured John Lennon appearing to throw the goat, leading to much speculation that he had the jump on Jinx Dawson and Jim Donlinger of Coven. Champions of the theory also point to the picture sleeve for the "Yellow Submarine" b/w "Eleanor Rigby" single, which features a photograph of Lennon flashing a goat-like sign. However, not only is Lennon's palm facing himself, but his thumb is extended outward—both faux pas, as any veteran goat thrower will attest. Most likely, Lennon is flashing the *Karana mudra*, a Hindu and Buddhist gesture to ward away evil.

The Beatle debate does, however, bring to light the utilization of goats by two of the Fabs' nonmetal contemporaries: The Beach Boys and the Rolling Stones. The Beach Boys' *Pet Sounds* (1966) famously featured sleeve photos taken by George Jerman and depicting the band feeding goats at the San Diego Zoo. Considered a magnum opus of popular music, the album was executed largely by Beach Boys founder Brian Wilson, with help from producer Phil Spector and a staff of ace session musicians known as the Wrecking Crew, while the rest of the band was on tour. Filled with impossibly sunny hooks and layered instrumentation, *Pet Sounds* influenced how the Beatles approached *Sgt. Pepper's Lonely Hearts Club Band*, according to Paul McCartney. The liner notes claimed the photo was an obvious play on the album's

The insert image of *Goats Head Soup* (1973), by the Rolling Stones, showed a goat's head staring up from a pot of soup.

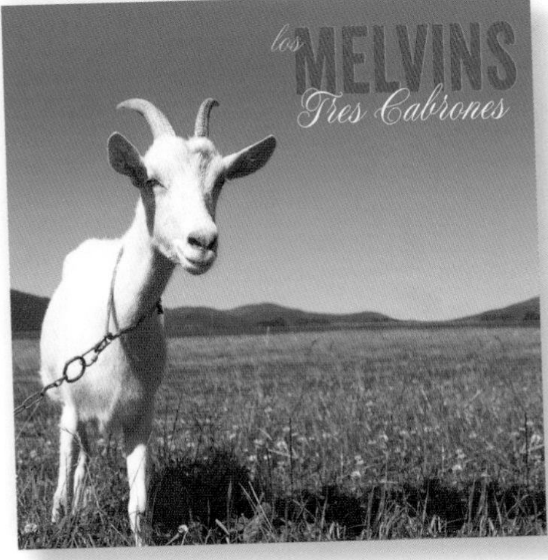

Grunge band the Melvins featured a dairy goat on the cover of their 2013 LP *Tres Cabrones*.

title, but guitarist Carl Wilson (Brian's younger brother), recalled that the title referred to the fact that the music was a collection of favorite, or pet, sounds. Whatever the case, the sleeve surely wouldn't be as memorable had Jerman brought the band to, say, the ostrich pen.

On the more diabolical end of the scale, far from sunny SoCal, are the Stones, with *Goats Head Soup*. Often unfairly maligned (following as it does *Exile on Main Street*, the band's general-consensus masterpiece), *Goats Head Soup* is best remembered by many for the album's insert image depicting a goat's disembodied head staring up from a bubbling cauldron. The album was largely recorded in Jamaica, where the title dish is said to be a delicacy. Whether that's true or not, the aforementioned insert, with the goat's vacant eyes staring at the listener, nicely complements menacing tracks about drug abuse ("Dancing with Mr. D") and police shootings ("Doo Doo Doo Doo Doo (Heartbreaker)").

Today the goat's relationship to music continues. No longer restricted to ancient troubadour traditions, opera, or heavy metal, the goat has even made inroads to punk (e.g., Aussie noise rockers Lubricated Goat), grunge (the Melvins titled their 2013 LP *Tres Cabrones*, which they translate to "three dumbasses" in their liner notes, and featured an inquisitive Saanen on the cover), to folk rock (the critically acclaimed Mountain Goats).

And then there are the Internet memes. A Google search of "the most heavy metal goat" pulls up 2.6 million pages featuring a four-horned black Jacob's goat (Brit rockers The Cult used the breed on the cover of their 1994 self-titled LP). The search term "punk rock goat" pulls up photos of a goat in the front row of what appears to be a basement club, intently watching the performance, while reactions from the band and fellow audience members range from mildly amused to totally unaware. This proves that, from ancient Greece to cyberspace, goats and music go together like rock and roll.

Unbreakable Record
THE 3 BILLYGOATS GRUFF
FRANK LUTHER
DECCA

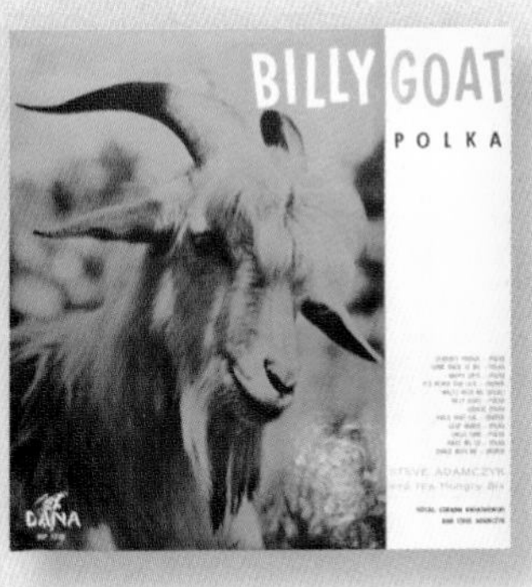
BILLY GOAT
POLKA

The Three Billy-Goats Gruff
A Norwegian Folk Tale

Darkthrone · Goatlord

The Electric Hellfire Club
"Kiss the Goat"

Alphonse Daudet
La chèvre de Monsieur Seguin
FERNANDEL
LETTRES DE MON MOULIN
DECCA

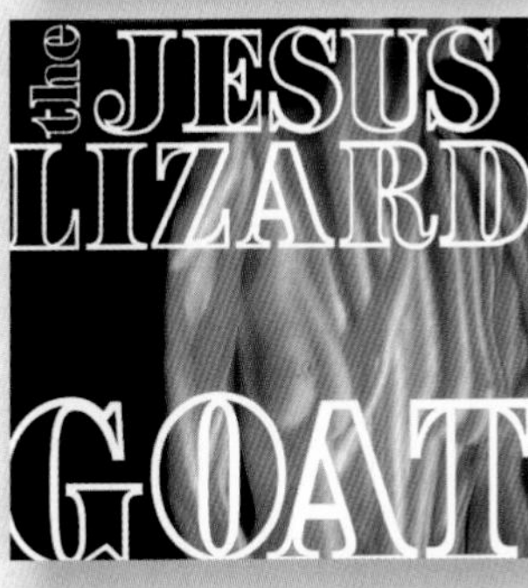
the JESUS LIZARD
GOAT

Inkubus Sukkubus
THE GOAT

Lubricated Goat
plays
the Devil's Music

MASTODON
The Hunter

Pentagram
Day of Reckoning

THE SHINS

Various Artists
NANNY GOAT SKANK

Show No Mercy
Slayer

THE CULT

Trees

wonderfuzzy
tree of goats

Classic evil goathead symbol behind the band in *This is Spinal Tap* (1984)

Here's a great one to know if you ever play Six Degrees with movies: Leslie Nielsen and Debbie Reynolds starred in *Tammy and the Bachelor* (1957)—with a goat.

Goats in Movies

By Elizabeth Noll

To paraphrase: now is not the best time to be a goat in movies. The brutal truth is that goats have never had the star power of other livestock—think chickens (*Chicken Run*), pigs (*Charlotte's Web*, *Babe*), or horses (*The Black Stallion*, *War Horse*)—or even spiders. That said, goats have lurked at the edges of the film industry since its inception. Charlie Chaplin (*Sunnyside*), Laurel and Hardy (*Angora Love*), and Buster Keaton (*The Goat*) were all aware of the goat's comedic potential. In the decades that followed, the goat's dark side was of more interest, referenced in such classics as *Dragnet*, *This is Spinal Tap*, and *Altered States*. A handful of recent movies, including *The Men Who Stare at Goats* and *Goats*, discard the weighty symbolism and re-introduce goat as animals—curious, amusing, curmudgeonly animals.

John Derek (B-movie star, husband of Bo, and director of *Bolero* and *Tarzan, the Ape Man*, among other films) sits on his dad's lap in a goat-drawn cart in Los Angeles (1927).

Movie poster of *Camping Out* (1919) with Fatty Arbuckle

Leni Riefenstahl (1902-2003), German filmmaker and actress, in Arnold Fanck's film *Der Grosse Sprung* (1927)

Mickey Rooney tries to drive a goat in the 1946 film *Love Laughs at Andy Hardy*.

Tom Hanks and Dan Akroyd (right of Hanks, in goat mask) on the job in *Dragnet* (1987)

Alexandra Paul (center) as The Virgin Connie Swail—in mortal danger—in *Dragnet* (1987)

A movie still from the comedy *The Men Who Stare at Goats* (2009), with George Clooney, Jeff Bridges, Kevin Spacey, and Ewan McGregor

George Clooney stares at a goat in *The Men Who Stare at Goats* (2009).

Gina Lollobrigida with a goat on the set of the 1956 version of *The Hunchback of Notre Dame*

Nancy and Ronald Reagan on their Malibu, California, ranch, with pet goats (1954)

Goats in Sports

By Steve Roth

There are four seconds left in a football game and the kicker fails to make an easy, chip-shot field goal that could have won the game.

It's the bottom of the ninth with two outs left and an infielder bobbles a routine ground ball that costs his team a run, and they lose the game.

What do these folks have in common? They would no doubt be called a *goat*. Sportswriters have been saying that for years to describe someone whose mistake or failure costs them the game. It's likely that "goat" is a simpler form of "scapegoat," which originated in the Old Testament and basically refers to someone who is singled out for blame.

It's one thing to call someone a goat; it's another thing altogether to blame an actual goat for a team's misfortune. If you're a Chicago Cubs fan, you know this has happened—and considering the Cubs' long record of futility, you might believe in The Curse of the Billy Goat.

In 1984 Sam Sianis, a nephew of Billy Sianis and current proprietor of the Billy Goat Tavern, brought a live goat out onto Wrigley Field in an attempt to lift his uncle's curse.

At the real Billy Goat Tavern, in Chicago, Illinois, a sign over the U-shaped counter reads "Enter at Your Own Risk." The sign pictured here is from the Saturday Night Live skit (starring John Belushi) that made the tavern famous across the country.

Wrigley Field

The Curse of the Billy Goat was supposedly placed on the Cubs back in 1945. According to lore, Billy Sianis, the owner of Chicago's famous Billy Goat Tavern (immortalized in the Saturday Night Live skit with John Belushi and Bill Murray: "cheeborger, cheeborger, cheeborger . . . no Pepsi, Coke!"), was asked to leave game four of the 1945 World Series at Wrigley Field because his pet goat's odor was bothering the fans. Sianis was so outraged at this offense he apparently declared the Cubs would never win another World Series. Or perhaps he said the Cubs would never win another World Series game at Wrigley Field. Or maybe he said that the Cubs would never appear in a World Series again. No one really knows for sure, though the Cubs *did* lose game four and, ultimately, the series. In fact, Sianis likely was at the game sans his olfactory-offending goat; an article in the *Chicago Sun* on October 7, 1945, says that Sianis's goat wasn't actually allowed into the game, and he left the goat tied to a stake in the parking lot.

Regardless of the veracity and accuracy of The Curse of the Billy Goat, superstitious Cub fans believed in it for decades. No doubt many still do, given that the Cubs haven't won a National League pennant since the incident, let alone a World Series, in 105 years.

Fans and the organization itself have tried to get rid of the curse, some in very novel and creative ways:

On Opening Day in 1984 and 1989, Sam Sianis, a nephew of Billy Sianis, brought a live goat out onto the field. In both years, the Cubs won their division.

Fans have tried to get into Wrigley and opposing teams' stadiums with goats in attempts to "lift the curse."

Sadly, a butchered goat was hung from the statue of legendary Cubs announcer Harry Caray in October 2007 when the Cubs were playing the Arizona Diamondbacks in the playoffs. That didn't work: the Cubs were eliminated by Arizona on October 6—the same date that Sianis brought his goat to Wrigley in 1945.

Possibly the best attempt at lifting the curse has also been the most beneficial. Reverse The Curse is a fundraising group in Chicago dedicated to bringing goats to families in poverty in developing countries. The goats provide these families with income from goats' milk, cheese, and meat.

Naturally, if you're reading this book you probably don't like the idea of blaming a goat for your favorite sports team's losses. Goats get into enough trouble all on their own. That's why when it comes to sporting endeavors, if we're going to mention a goat at all, let's use it as an acronym: G.O.A.T. = Greatest Of All Time.

Goat drinking mugs of beer. April 1978.

Goats and Beer

By Dennis Pernu

It seems inevitable that the occasionally affable (but more often surly) billy goat would somehow become associated with beer, that amber elixir known to make human moods similarly oscillate. Indeed, for nearly four hundred years, the billy goat has been the de facto face of bock, a type of lager that originated in Germany but which inevitably found its way across the Atlantic. But it's not beer's transformative qualities that made the billy goat a no-brainer for this particular pitch job.

Around six hundred years ago in central Germany, brewers in the town of Einbeck—perhaps trying to create something new under the restrictive *Reinheitsgebot*, the so-called "German Purity Law" that dictated that brewers could use only malt, hops, and water—hit upon a malty, lightly hopped dark ale. Three centuries later and farther to the south, Munich brewers adapted the style to a lager. Giving props to their neighbors to the north, but not quite able to correctly pronounce the name "Einbeck," the Bavarians referred to this new style of beer as *ein bock*—literally "a billy goat" in German. As Charlie Papazian notes in his landmark work, *The New Complete Joy of Homebrewing*, the typical bock "is strong in alcohol with a very malty-sweet overall character. Hop bitterness is low and only suggests itself to offset the sweetness of malt."

Unlike ales, lagers are fermented at relatively cold temperatures, typically below 45 degrees Fahrenheit (7 degrees Celsius), and for longer periods than ales. In the nineteenth century, during the United States' first beer boom, immigrants from the European mainland brought lagers with them, which eventually became the go-to style for the hundreds of small regional brewers (including a pair of small upstarts in the Midwest calling themselves Budweiser and Miller). As the number of breweries inevitably contracted over the decades, lagers remained and were most likely the beers your father or granddad stashed in that old Kelvinator out in the garage.

Though many are produced year-round, bocks are often associated with seasons in the United States—lighter-bodied *maibocks* release in the spring, and the darker bocks (with higher alcohol content) make nice seasonal warmers in the autumn and winter. Papazian even notes that some "Christmas bocks," appropriately enough, are intended to be consumed under the sign of Capricorn.

Before American brewers streamlined their offerings, focusing on a single, flagship beer (most often a straight-up yellow lager), bocks were standard seasonal offerings from many brewers of the nineteenth and early to mid-twentieth centuries. For collectors of breweriana, this means there's no shortage of labels, cans, and posters from across the country depicting billy goats in all states of comportment and apparent inebriation. Stately or indignant, suave or uncouth, genial or belligerent, the billy goats found on the artwork from this segment of collectibles are fantastic, ranging from colorful, pastoral, Victorian scenes to mid-century caricatures worthy of the most sophisticated Midtown ad agency.

Beer stein, ca. 1982

Poster for C. Feigenspan's bock beer, ca 1880-1890

Today, amid America's exploding craft beer craze, scads of ales and lagers of all stripes incorporate goats into their names and label art. At Dave's Brew Farm, a wind-powered operation in tiny Wilson, Wisconsin, beer enthusiasts can duck into the taproom for a pint of Matacabras, a dark Belgian ale whose Spanish name translates to "wind that kills goats."

"[W]hen we stumbled upon the name of this wicked Spanish breeze," the staff explains, "we decided to brew up something that would likewise stop folks in their tracks."

In upstate New York, Two Goats Brewing is named for founder and head brewer Jon Rodgers' love of a good doppelbock (i.e., "double bock"). Jon and his wife Jessica carry the theme through a number of their acclaimed beers, including Redbeard Red Ale, Dirty Shepherd Brown Ale, and Danger Goat! Blonde Doppelbock.

So the next time you're hoisting a brew, toast the ubiquitous billy goat and his inspiring visage. Without him, the world of beer just wouldn't be the same.

FRANK FEHR
BREWING CO.
(INCORPORATED)
LOUISVILLE, KY.
TRY
OUR
BOCK BEER

Eichler's
Bock
BEER

BOCK
BEER
BOCK BEER

CONTENTS
12 FLUID OZ.
Fort Pitt
BOCK
Beer

CONTENTS 12 FL. OZ.
Belmont
HIGH PROOF
DOES NOT CONTAIN MORE THAN 4 PER CENTUM OF ALCOHOL BY VOLUME
BOCK BEER
BELMONT BREWING COMPANY
MARTINS FERRY, OHIO

ERIN
BREW
BOCK BEER
THE STANDARD BREWING CO., CLEVELAND, OHIO

CARLING
CARLING
Black Label
BOCK
BEER

DUQUESNE
BOCK BEER

CONTENTS ONE FL. QUART
Burger
BOCK
BEER
REGISTERED

G.F. BURKHARDT'S
BOCK BEER

SAMUEL ADAMS
SAMUEL ADAMS
CINDER
BOCK
RAUCH BOCK

FEIGENSPAN'S
NEWARK
N.J.
BOCK BEER
CONTENTS 12 FLUID OUNCES
Cook's
BOCK
BEER

Du Bois
BOCK
BEER

FOX
DE LUXE
Bock
Beer

Champagne Velvet
Brand
BOCK BEER
CONTENTS 12 FLUID OUNCES
ALCOHOLIC CONTENT IN EXCESS OF 3.2% BUT NOT OVER 7% BY WEIGHT

FRANKENMUTH
BOCK BEER
FRANKENMUTH BREWING CO.

Gettelman
BOCK
BEER
Brewed in the True
Milwaukee Tradition

Genuine
Draft Beer
Einbock
Draft
BOCK
©1966 NET CONTENTS 1 QUART

12 FLUID OZS.
HEURICH
BOCK
BEER
CHR. HEURICH BREWING CO.
WASHINGTON, D.C.
INTERNAL REVENUE TAX PAID

CONTENTS 12 FL.OZ.
Koch's
BOCK BEER
BREWED AND BOTTLED BY
FRED KOCH BREWERY, DUNKIRK, N.Y.

BOCK BEER
BREWED AND BOTTLED BY
KIEWEL BREWING CO.
LITTLE FALLS, MINN.

Gunther's
BOCK BEER

Bock Beer
Huber

G
B
DOUBLE BOCK
BEER

Knapp's
Bock
Beer

CONTENTS
12 FL.OZS.
KOEHLER
GENUINE
Bock
Beer

HoLiDay
BOCK
BEER
A Genuine
Old Fashioned

HENSLER
BOCK BEER

CAMDEN
CONTENTS
1 QUART
BOCK BEER
BREWED BY CAMDEN COUNTY BEVERAGE CO. CAMDEN, N.J.

KRUEGER
K
BOCK
BEER

Ballantine
Bock Beer

KAMM'S
BOCK
BEER
CONTENTS 12 FLUID OUNCES
BREWED AND BOTTLED BY
KAMM & SCHELLINGER CO., INC.

Knickerbocker
BOCK
New York's Famous Beer

Seitz
BOCK
BEER

BAMBERGER
ORIGINAL
Mahr's
Der Weisse Bock
MALT LIQUOR
1 PT .9 FL OZ

Old Crown
LAZY-AGED FOR MONTHS
BOCK
Beer
IN BOTTLES AND CANS
100% employee owned and operated

BrewFarm
MATACABRAS
A CURIOUS ALE

F. KLEMM'S
BOCK

Ortliebs
LAGER BEER
BOCK ALE
MARKET 4728
HENRY F. ORTLIEB
PHILADELPHIA
MADE IN U.S.A.

CONTENTS 12 FL. OZ.
HITS the SPOT!
PEOPLES
BOCK BEER
Brewed to please YOU!
BREWED AND BOTTLED BY THE PEOPLES BREWING CO., OSHKOSH, WIS.

old
BOHEMIAN
BOCK
BEER
FULLY
AGED
EASTERN BEVERAGE CORP. — HAMMONTON, N. J.
FULL QUART

PABST
GENUINE
BOCK
BEER

Old German
BRAND
BOCK BEER
GERMAN BREWING Co.
CUMBERLAND, MD.
NET CONTENTS 12 FL. OZ.

Pabst
1913
Bock Beer
Rich–Smooth–Delicious
A pure malt beer of extra strength and quality. In this brew we again make good our claims of presenting to the public the finest glass of beer ever produced.
On Draught or in Bottles
PABST BREWING COMPANY
Order a case for your home
Phone Grand 5400

PARK ROW
Bock Beer
CONTENTS ONE QUART
PARK ROW BREWING CO., NEW HAVEN, CONN.

OLDENBERG
OUTRAGE-O-US
BOCK
MICROBREWED BOCK BEER

LONE STAR
BOCK
Brewed in Texas
The NATIONAL BEER OF TEXAS

BAMBERGER
ORIGINAL
Mahr's
Christmas Bock

First Brewed in 1885
NATIONAL
BOHEMIAN
Bock Beer

THE ORIGINAL
OLD GERMAN
Beer on Record
BOCK BEER

Schell's
BOCK
THE AUGUST SCHELL BREWING COMPANY · NEW ULM, MINN
AUGUST SCHELL
SINCE 1860

INTERNAL REVENUE TAX PAID
CONTENTS 12 FLUID OUNCES
BOCK
BEER
BREWED AND BOTTLED BY
WEBER WAUKESHA BREWING CO.
WAUKESHA, WIS.

Tivoli
Bock Beer

Stanton Giant
BOCK
SINCE 1817
BEER
Absolutely the Best ever Brewed

Pfeiffer's
Bock Beer
CONTENTS 12 FLUID OZ.
Pfeiffer Brewing Co.
ESTABLISHED 1889

Schaefer
BOCK
THE F. & M. SCHAEFER BREWING CO.
NEW YORK, N.Y.
PERMIT NO. U-215

Renner
BOCK
BEER
THE RENNER COMPANY
YOUNGSTOWN, OHIO

SKOVLYST
Limited Edition
DoppelBock

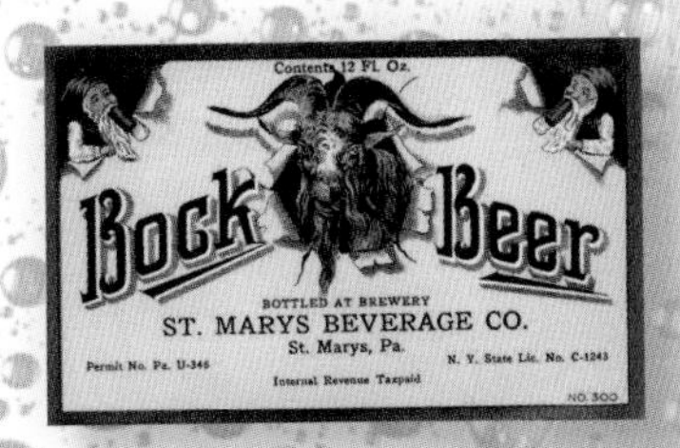
Contents 12 FL. Oz.
Bock Beer
BOTTLED AT BREWERY
ST. MARYS BEVERAGE CO.
St. Marys, Pa.
Internal Revenue Taxpaid

Schoenling
OLD TIME
BOCK BEER

SPOETZL
BREWERY
ENJOY
Shiner
BOCK
BEER
Prosit!

CONTENTS 12 FLUID OUNCES
Walter's
BOCK
Beer
BREWED AND FILLED BY THE WALTER

DOUBLE BOCK BEER
CONTENTS 12 FLUID OZ.
TIVOLI BREWING COMPANY · DETROIT, MICHIGAN

TWO GOATS
BREWING
HECTOR NY

EST'D 1860
Schmidt's
OF PHILADELPHIA
BOCK
BEER

Old
Bohemian
BOCK
BEER
FULLY AGED
Brewed with Pure Artesian Well Water
CONTENTS ONE PINT

Straub
BOCK BEER
Brewed with pure Mountain Spring Water

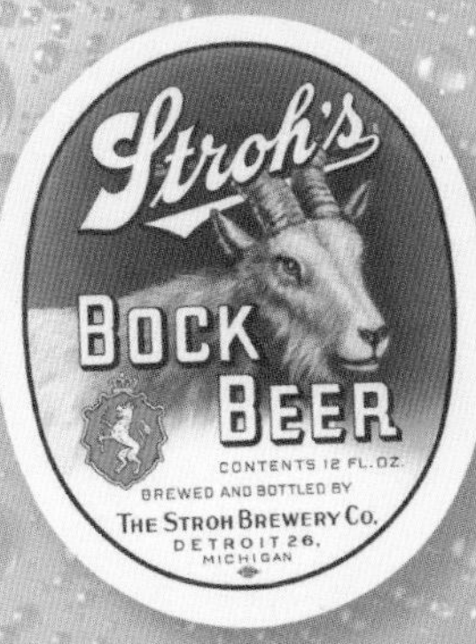
Stroh's
BOCK
BEER
CONTENTS 12 FL. OZ.
BREWED AND BOTTLED BY
THE STROH BREWERY CO.
DETROIT 26,
MICHIGAN

Yuengling
SINCE 1829
BOCK
BEER
MADE AND BOTTLED BY
D.G. YUENGLING & SON, Inc.
POTTSVILLE, PA.

BOCK BEER
BREWED AND BOTTLED BY
LOUIS ZIEGLER BREWING CO.
BEAVER DAM
WISCONSIN

5 ★ GOATS AT WORK

By Janet Hurst

A goat is more than just a pretty face. They are hard-working creatures, and not just as meat and milk producers; they can be trained to clear the landscape of invasive plants, to pull small carts and wagons, to pack gear along back country trails, and many other jobs.

The city of Mesa, Arizona, uses goats to remove invasive trees and weeds from the grounds at the water reclamation plant.

Landscape Management

Recently goats received national attention for their role as groundskeepers for the Congressional Cemetery in Washington, D.C. To combat the growth of invasive plant species in the cemetery, these four-legged lawn mowers were introduced to provide brush and weed control. They quietly go about their daily task of consuming unwanted weeds, providing a needed service for a reasonable cost.

Goats have been used extensively in conservation practices. In areas where the forest is an overgrown tinder box, the goats can clear away low-growing shrubs and eat dry leaves and other plant material. This type of mitigation reduces the likelihood of fire in areas where the animals are allowed to forage. Goats can go into ravines, down hillsides, and into other areas difficult for humans to reach. Services are available for those who don't own their own herd. Such companies as Rent A Goat, Rent A Ruminant, and others provide a goat rental service for a fee.

Owner Brian Knox calls his herd "Eco-Goats." They can clear an acre of invasive species (including poison ivy and thorny multiflora roses) in one week or less.

Airports are among those companies utilizing goats as lawnmowers and brush control agents. The use of these animals eliminates the cost and environmental problems associated with chemical means of weed control. The goats are moved across the acreage in a series of portable paddocks, controlling their range and encouraging them to clear one area at a time. Goats are currently on post at Chicago O'Hare, San Francisco, Atlanta, and several other airports. The Extension (www.extension.org) website reports: "For well over 100 years, goats have been used to manage unwanted vegetation in the United States. The role of goats in vegetation management is expected to expand in the near future due to a combination of factors, including environmental degradation created by past systems of farming; reduced efforts to control unwanted vegetation that has increased its spread and growth; the increasing expense of mechanical methods of vegetation control, most of which rely on the use of fossil fuel-driven machinery; and the increasing reluctance and moral unacceptability of the use of herbicides due to pollution hazards and potential long-term harm to the environment."

The Pacific Northwest was the first region to adopt goats as brush control agents in urban areas. The projects have been so successful that the Seattle City Light Department and the Seattle Parks and Recreation Department now rent goat herds on a regular basis. The city of Chattanooga, Tennessee, opened its doors to a small herd of goats with the intention of clearing kudzu (an extremely invasive plant species), with outstanding results. The animals are hard at work on sections of the Appalachian Trail, clearing peaks of brush and brambles.

At the Chesapeake Montessori School in Annapolis, Maryland, a goat helps maintain the grounds, stepping in where weed-whackers and herbicides have failed.

In the lush Pacific Northwest, escaped garden plants such as English ivy and rhododendrons have become invasive, outcompeting native plants and growing out of control. Goats, however, make short work of it all, as on this hillside in Portland, Oregon.

Goat wagon in Eden Park, Cincinnati, Ohio, ca. 1906

Goatman Charles Ches McCartney poses next to his goat-drawn wagon full of his wares; United States, ca. 1920s.

Draft Animals

Goats are often trained to pull small carts and plows, serving as miniature draft animals. Goat carts have been specially designed for centuries for the animals to pull small loads of grain, hay, or other cargo. These animals can pull about 1½ times their own body weight. The agile, sure-footed goat has no problem heading up or down a trail. Goats also can be taught commands using the tried and true directions of *gee* (right), *haw* (left), *giddyup*, and *whoa*. Repetition and consistency are the keys in training all animals.

According to the Harness Goat Society, nineteenth-century German goat carts were used in poor farming communities to bring produce to market. Because goats are smaller, they require less feed, and for small farms, it didn't matter that they couldn't haul as much as a horse. In addition to hauling produce, goat carts were used to carry children and disabled people.

This young boy may have been disabled and dependent on goat power to pull his wheeled chair.

Goat packing in the Red Desert of Wyoming

Beasts of Burden

Perhaps one of the most intriguing uses for a goat is as a pack animal. Pack goats are used in mountainous regions or on open trails, wherever a beast of burden can be advantageous. There are companies who deal in pack goat gear, offering miniature-size panniers, saddles, and other goat accoutrements.

Due to the fact that goats are herd animals, they seek companionship. That can come in the form of another goat, another animal species, or from humans. This makes them perfect for hiking, as they will literally follow the leader. A well-muscled goat can carry 20 to 30 percent of its body weight. Larger breeds are most appropriate: Nubians, Saanens, Oberhasli, Toggenburgs, and Alpines are good choices for pack goat training. Training begins when the animal is quite young, with small packs introduced at first. The animal becomes accustomed to carrying larger and larger packs as it grows. Some trainers specialize in working with pack goats and breeders who sell goats with skills developed specifically for packing.

Why a goat as a pack animal? There are several reasons. Cost is one. Goats, even trained animals, are less expensive than horses. They are more compact than mules. They eat brush, making feed easy to obtain. They enjoy human companionship. A goat can even provide milk on the trail. A goat train of several animals is not an uncommon site in mountainous terrain. The animals can be tethered together to add carrying capacity. Noted pack-goat enthusiast John Mionczynski is credited with the introduction of pack goats to the United States in the 1980s. His book *The Pack Goat* is a classic. The North American Pack Goat Association was formed in 2000. Goats for packing are found mostly in the Pacific Northwest, in Washington at the Olympic National Forest, and in the Columbia River Gorge areas. California also offers pack goats at the Armstrong Redwoods State Reserve, Annadel State Park, and other areas in Sonoma County. Additionally, pack goat outfitters are found in Wyoming, Idaho, and Utah. "Goat hikes" have become a popular event in Maine. As this trend catches on, expect to see more goats on the trail.

Caprine Therapy

Donkeys, miniature horses, and goats have been used in animal-assisted therapy. Their winning personalities make them a natural fit with children and the aged. Studies have shown that goats, dogs, and other therapy animals can ease feelings of separation from loved ones; lower blood pressure, heart rate, and stress levels; and reduce aggression in people with Alzheimer's. Goats are popular nursing home visitors, with many residents having a goat tale or two to share. The animals have been shown to provide physical and emotional therapy to those who interact with them. On the other end of the age spectrum are children, who cannot resist a goat. Interaction between goats and autistic children has proven successful. Goats always provide a laugh and a smile, which is good medicine at any age.

Caprine therapy? Sure. The presence of animals such as dogs and goats lowers stress levels and often brings a smile to those who need it most.

6 ★ A GOAT HERD

Reprinted with the permission of Scribner Publishing Group from *Goat Song: A Seasonal Life, A Short History of Herding, and the Art of Making Cheese* by Brad Kessler.

Before I raised goats I entered the woods to hike or hunt; find birds, mushrooms, solitude, or cordwood. Yet with the goats I went into the woods with no purpose other than to herd. I brought along a walking stick, a pocketknife, a bell. I brought a hat, a flannel shirt for a pillow. I once brought a rifle but realized—all wrong. Sometimes I brought a book. Always pen and paper.

The walking stick was not for herding goats (they wouldn't respond to a stick). I used it to bend down branches or knock crabapples off their boughs. The bell was for calling the goats. Mine was a hammered copper on a leather thong that I tied to a belt loop. The goats wore bells that came from around the world—bronze or copper, iron or brass; some had wooden clappers or washers hung on rusted wires. We fashioned collars from old men's belts augured with new holes. Each doe wore her own belted bell so we could tell who was lagging or missing by the tone. Hannah, the queen, sported the largest bell (made of brass), the

Domestic goats depend on their human to herd them safely through the forest and back home.

Goats prefer to be in open spaces where they can see for miles; that way they can see their predators approaching.

kids tiny copper ones no larger than matchboxes. All the bells had different timbres and together made an atonal music—a tintinnabulation—those afternoons. We sounded like a herd of North Pole reindeer.

The women herders of Sweden, the *valkulla*, once clanged pots and pans on their first forays into the woods to frighten off bears and wolves. I rang my bell each time we entered the forest. I rang it not for the goats to follow (they would anyhow) but to alert whoever else was around: coyote, fisher, bobcat, fox, bear, and that other predatory animal: humans.

The pocketknife was for slicing apples, or cutting free leather collars should two does get their heads stuck in one, as happened once when Nisa and Pie reached for a pile of acorns and came up with the same belt garroting both their necks. They nearly choked themselves fighting to get free. I had to cut through the hard leather and throw away the straps.

Herding is a menial task but also, historically, a prophetic calling. The shepherd-king reaches back long before Christ—the Good Shepherd—and predates all the herder-prophets of the Old Testament: Moses and Joseph, David and Amos. Perhaps the motif arose with Dammuzi, the Sumerian shepherd god, or maybe with Krishna, the Hindu shepherd god. Wherever it began, the image of the hero who leads his flocks in the fields and protects them from the wild lies buried in ancient human phylogeny.

The poetic muses came to Hesiod while he tended flocks on Mount Helicon. Moses walked his goats and sheep in Midian when the Burning Bush spoke to him. Shepherds first received word of Christ's birth; and the shepherd Muhammad heard the words of God in the wilderness. *No man becomes a prophet*, he later wrote, *who was not first a shepherd*.

Herding is an impoverished occupation, a job often relegated to children, the elderly, the slave or exile (Saint Patrick was a shepherd slave). Today the poorest people in the most impoverished nations subsist as herders of sheep and goats. Why did God speak to shepherds and not, say, cobblers or dry-goods salesmen? Was it because shepherds lived apart, on land too dry, steep, or poor for other humans; or because they had all day to both dream and stay awake, and their hours were spent alone, in silence, watching animals, weather, and wind; and their minds could wander like their sheep and goats?

As long as there were shepherds, they migrated with the seasons. In summer they brought their herds to the high country where the grass was coming green. In winter they brought them back to the lowlands or river valleys to pasture them or feed them stored fodder. The system of moving animals seasonally to better grazing is called "transhumance," from *trans* for "across" and *humus* for "soil." *Across the soil.* Transhumance is ancient; the animal tracks it followed date back in some places to prehistoric times. Transhumance still occurs today in almost every continent on earth. In the Swiss Alps it's called the *Alpaufzug*, in the Chilean Andes, the *veranada*, in the Swedish north, the *buforing*. In the Hindu Kush, in Kenya, in the Pamirs and Peru, the animals go up the valleys in spring like rivers running in reverse. All summer the herds stay up in the high pastures—the *alpage*, the *seter*, the *jailoo*, the *estive*, and *son*. It's often a bleak and lonely business for the herders. The days last long and the work is tough. But every account of the transhumance, from today's

Herders in Myanmar (Burma) walk with their goats

Spanish herders with their goats in Andalusia

Herders from every culture, on every continent, find magic in the time spent alone with their animals during the summer months.

A leisurely stroll through woods or meadows gives goats a chance to nibble wild greens and other delicacies.

Pakistan to the nineteenth-century Scottish sheilings, mentions the beauty and magic of living on mountains with herds. The Basque shepherds (all men)—who last century bunked together during their summer transhumance—bragged about the splendor of the Pyrenees and the purity of the mountain cheese they made up there. The Swedish herders (all women)—who last century walked their cows and goats to the remote spruce forests each summer—developed special songs for their herds. The women used a vocal technique called *kulning* that their animals were said to understand. When the *valkulla* returned to their villages in the fall, they hung up their herding bells and refused to sing their *kulning* songs back home. *Kulning* was the language of their summer transhumance, one meant for communion with other animals, and not with men.

On my daily walks in Vermont I tried to incorporate the wisdom of transhumance. Each day I left the worn-out low-lying pastures and headed for higher ground. We walked away from home—from parasites and manure—and sought wilder earth. A lactating doe requires about eight pounds of dry matter a day. On our daily walks my goats ate far less than that, but the woods gave them a little of what they needed—wild greens rich in minerals and carotene. The walks gave me what I needed too.

Only a few decades ago—before they shipped their cattle on trucks—the transhumant herders of the high Spanish sierra walked their Avileña cattle down to the plains of Estremadura in early spring. The journey took several days, and hired herders did the job, yet the village men always went along too. When an anthropologist asked the men of one village why they walked their herds if they didn't have to, they made all sorts of excuses (the cattle were nervous; the herders needed help). Finally they all confessed: they walked their cattle because "we like doing it."

We like doing it. And I liked herding my goats, and people in cities like walking their Yorkies and pugs. There's something in the rhythm of walking animals—*andante, lento, corrente*—the companionable pace, the striking out into the world, the earth alive around us as we walk.

7 ★ GOAT PORTRAITS

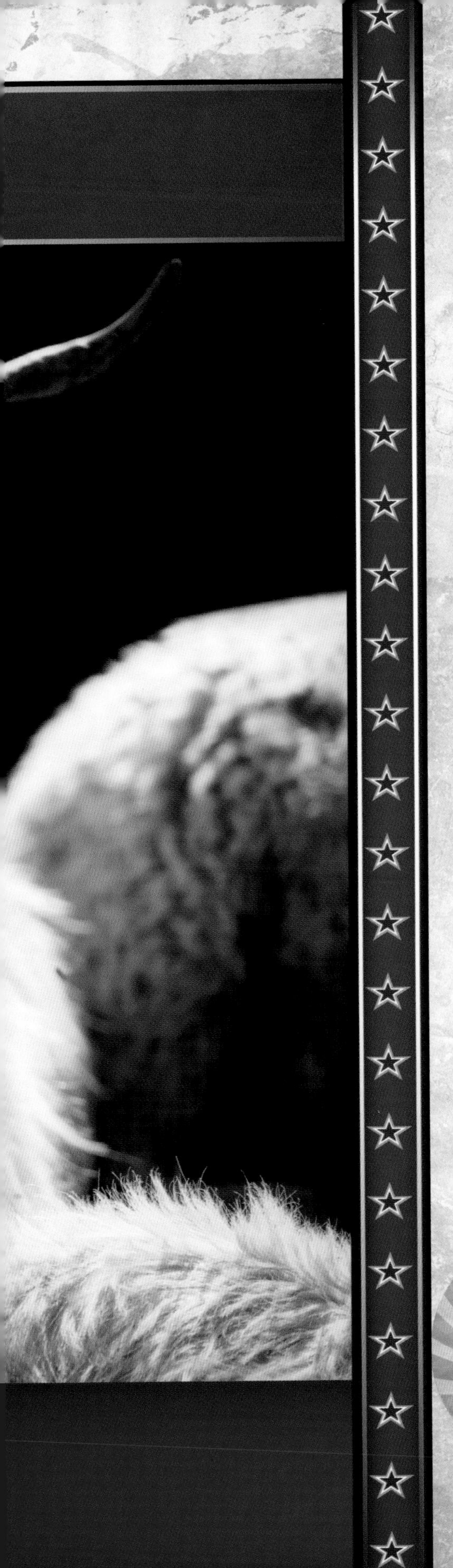

It has been said that goats have more personality than any other farm animal. They have minds of their own, they get attached to their owners, and they're amusing and endearing—that is, unless they're bent on revenge or up to some mischief. These portraits reveal the surprising range of the caprine character, from grave to silly, from delicate to dignified.

8 ★ THE FIBER OF GOATS

Ask ten people where cashmere comes from: nine won't know it comes from goats.

Fiber animals are some of the most beautiful beasts to inhabit the earth. Sheep, llamas, alpacas, Angora rabbits, Angora goats, and Cashmere goats give us luxurious fabric. A mature Angora goat will produce ten to fifteen pounds of mohair each year, making it an ideal addition to a fiber-producing herd.

Mohair

Angora goats give us mohair. They are known for their curly, silken locks and angelic appearance. This breed is typically docile, most concerned with browsing and producing hair.

The fiber from Angora goats has a shimmery, silken texture known for its durability. Military uniforms of the past, high-end upholstery, and even covering for airplane seats have been made of fabric containing mohair. Spinners and weavers value this fiber, making the kid curls into a spectacular yarn. When spun with a trained hand, the tiny curls create slubs and extensions in the yarn that then are highlighted in the crafting of a garment. The tiny curls remain visible in the yarn, producing a texture that is unsurpassed. Manmade fibers cannot duplicate the delicacy of this finely spun yarn. Conversely, the yarn can also be spun into thick, ropelike material. Highly durable, this type of yarn can be used for projects such as purses or bags that will receive heavy use. Doll wigs and Santa beards are often made of mohair, too. If you are lucky enough to find an antique Santa suit, chances are the beard will be made of luxurious mohair locks. Often the trim on the suit will be made of fabric containing mohair.

With all these curls, Angoras have a tendency to pick up debris such as cocklebursʼ, small sticks, and other materials in their locks. Some farmers who raise fiber goats will put a lightweight coat over them to keep their hair clean. Maintenance of their browsing area can prevent some debris collection, but it is largely inevitable.

Angoras seem to be less hardy animals when compared to their dairy and meat goat cousins. It's important to provide special care at birthing time, shelter during harsh weather, and frequent health checks.

Reprinted with permission from *The Whole Goat Handbook* by Janet Hurst (Voyageur Press, 2013)

Angora goats are good mothers, and their kids, with shimmery, silky, bright white curls, are adorable.

The reproductive cycle of the fiber breeds is the same as that of the dairy breeds, with gestation occurring 145 to 150 days after breeding. Twins are less frequent than in dairy herds.

In my personal experience, I have found Angoras to be attentive mothers, with rejections rare. The kids are the most precious creatures I have ever seen. They are bouncy little balls of curls with eyelashes like those of a china doll. These animals are the cure for the midwinter blues, and their births are something I greatly look forward to each year. Baby goats are truly miniature versions of their adult counterparts: tiny little hooves, little knobs that will become horns, and those tight curls that make them appear they have just had a curly perm. Irresistible!

Shearing

Angora goats are sheared twice a year—once in early spring and once in the fall. Watching the weather is important, because without their coats, the goats will chill if a cold snap happens unexpectedly. I keep a stack of children's sweatshirts with the sleeves cut off in the event a kid becomes chilled. Of course my neighbors talk about that crazy goat lady with her goats in sweatshirts; however, this effort has proven to be a viable means to warm up a chilled goat and reduce the risk of loss. Commercially made goat coats are also available.

Shearing is an acquired skill, and it is best to go to a shearing school or to hire a skilled shearer to take care of this task. Without proper training and equipment, unskilled shearers can easily harm the animal with the blade of the shears. Mohair does not contain lanolin, as wool does, so goats are more difficult to shear. It takes time and practice to learn this skill and perform it with certainty.

One of the worst things that can happen during shearing is to shear off a nipple. This unfortunate mishap causes problems down the line when the goat becomes pregnant again. There is no way to relieve the pressure of the milk, which will still be produced

This is mohair sheared from an Angora goat. White is the most common color and ranges from true white to ivory. The silvery gray is called "blue" in the language of goat owners.

and present in the udder. Most good shearers will cover the teats with their fingers while working the belly to avoid this disaster.

General cuts from the sharp blades are also a risk, and it is not uncommon to have to stitch up a wound. Blood-stop powder is available at farm supply stores and should be kept on hand for shearing day and other accidents. This substance coagulates the blood quickly and helps to stop any bleeding.

Wool carders are an investment, but they'll last for years. A carder arranges all the fibers in the same direction. Whether you're spinning or creating batts for felting, carding is a necessary step. Purchase new carders for best results; antique carders may have old, rusted wires.

Collecting Mohair

Shearing twice a year allows the Angora coat to grow to a (staple) length suitable for carding. Carding may be done manually using wool carders or sent to a carding mill. First, the fiber should be manually picked through to remove any debris or vegetation. It is important to keep the curls as intact as possible.

Currently mohair is priced at three to twelve dollars a pound. The great variance is attributed to the length of the staple; whether the mohair has been skirted (vegetation removed), cleaned, or washed; and whether the fiber is from a kid or a mature animal. Kid hair brings the highest price. Naturally colored mohair is priced at ten to twenty-five dollars a pound with the same variances applied.

Older goats lose the tightness of the curl, and their fiber does not have the same sheen or texture. It will develop "kemp," which is a far less desirable fiber. Kemp is somewhat matted, and while it can be carded and spun, it does not have the attributes associated with mohair.

Mohair wool that's been cleaned, carded, and dyed

Another term used regarding the fiber is "stain." Mohair from the goat's rear quarters often has a dark stain. Many spinners capitalize on this natural coloring and do not dye this hair but leave it as is. After washing, a reddish color will remain, and the color variation is unique.

Separate the stain from the rest of the fiber, and set it aside for special projects requiring colored hair. After picking out the debris, wash the hair.

This procedure must be done with care. If the hair is agitated during washing, it can felt, which results in a matted clump. Place the locks in a nylon bag (a laundry bag or lingerie bag), and then soak them in a lidded container of hot (145° F) water. Add a bit of dishwashing detergent or fiber wash to the water. Lay the bag in the water, pushing it in to submerge it. Let it soak for a few minutes, and then turn the bag over and let it soak again.

After forty-five minutes have passed, take a lock out of the bag to see if it feels clean or if it is still waxy. If it is waxy, remove the bag from the container, drain the water, and then add more hot water and detergent. Repeat the soaking until the fiber feels clean. When you are satisfied with the results, keep all the fiber in the bag and put it into your washing machine on the spin cycle. Do not add water; you are simply spinning out the excess water in the bag. When this step is finished, remove the mohair from the bag and place it on a screen to dry. When it is completely dry, it is ready to be carded. Some spinners will tease the locks a bit, which will add more texture to the yarn. To store fiber that has been washed but not yet carded, cloth bags are best. Suspend the bags, and keep them in a dry place where insects or mice will not have access.

Carding arranges all the fiber so that it all goes in the same direction. Fibers are worked on carders that resemble dog brushes. The metal teeth of the carders bite into the mohair (wool, cashmere, or other fiber) and align them for spinning.

To card using hand carders, arrange the locks on the carders and simply work one against the other, pulling the locks so the fibers all go in the same direction. To preserve the curl, pull lightly (some spinners work directly from the lock and avoid the carding process). Peel the prepared hair off the carders, and roll it in to a tubelike shape called a "rolag." Store the rolags in a basket, and set them aside for spinning.

A drum carder will automate the carding process. Some models have a hand crank and are not electrified. Others have a small electric motor that takes much of the work out of this step. Carding mills will process the fibers for a fee and send them back nice and clean, ready to spin. Fibers are returned as roving ready to be spun, premade yarns, or batts for felting.

Cashmere

Cashmere often holds the claim to the most luxurious fiber on the planet. Amazingly, few actually associate the origin of this luxury with a goat. Those who consider goats to be revolting creatures will proudly sport a cashmere sweater or scarf, remaining oblivious to the producer of the silken fibers. It is said the best-quality fiber is that which the goat actually sheds. This fiber is gathered by hand. Combing is another option for the small herd. However, most commercial operations today shear their animals annually rather than go on a scavenger hunt for the fibers left in the pasture. This fiber is treasured for its texture and sheen. With an adult producing about four ounces per year,

Because cashmere is of such high quality, it can even be recycled. These two hanks of cashmere yarn were made by unraveling old cashmere sweaters and then washing and respinning the yarn.

it is small wonder that a cashmere garment is such an investment.

Cashmere wool fiber is obtained from Cashmere and other goat breeds. The goats produce a double fleece that consists of a fine, soft undercoat, or underdown, of hair mingled with a straighter and much coarser outer coating of hair called "guard hair." For the fine underdown to be sold and processed further, it must be de-haired. De-hairing is a mechanical process that separates the coarse hairs from the fine hairs. After de-hairing, the resulting cashmere is ready to be dyed and converted into yarn, fabrics, and garments.

An interesting product is now available: recycled cashmere. Due to the fact that cashmere and mohair create such long-lasting upholstery fabric or garments, it is possible to find vintage pieces in the local thrift store. Fiber enthusiasts seek out old cashmere sweaters and other garments, clean and unravel them, and then re-spin them (often combined with another fiber) to create beautiful new skeins of yarn.

Dyeing

Mohair takes dye exceptionally well. The colors remain vibrant and intense. The fiber may be dyed before or after spinning. Accents of the fiber will add color and texture to weavings and knitted or crocheted items. Beautiful, naturally colored fibers are available as well. Selective breeding has brought about a fabulous array ranging from reds and blacks to silver and rich grays.

Fiber may be dyed at several stages after the initial cleaning. Dry fiber may be added to a dye bath before carding, creating lovely variances in intensity. Wool may be spun and then dyed in a skein. Mohair particularly picks up dye with trueness to the intended color, producing a vibrancy and intensity not found with other fibers, and dyeing enhances the natural sheen of mohair. Experiment with types of dye. You'll find a huge variety of specialty products intended for the spinner. Some have varying shades of color built into them, giving a gradation of color within the fiber. Natural dyeing is a science unto itself.

Carol Leigh of Hillcreek Fiber Studio works at the dye pot. Hues vary from light to dark depending on the amount of time they are soaked in the dye.

These naturally dyed yarns have been gathered in hanks and hung to dry.

The drop spindle is an inexpensive means of spinning fiber into yarn. If you are new to spinning and simply want to practice, a drop spindle will serve you well.

Yarn Painting

Yarn painting (or roving) is one method of adding color. This technique works well with long pieces of roving (eight feet long or so) or skeins of yarn. To paint roving, begin by tying lengths of roving into a bundle. Place the bundles in hot water with a squish of dish soap. Let the roving sit in the hot water for about an hour. Push the bundles into the water, but do not agitate them (we don't want felt). After soaking, drain the water and push the bundles lightly to remove excess water.

Cover a work table with plastic. Next lay down lengths of plastic wrap long enough to accommodate the wet roving. Tape the ends of the wrap in place. Mix up your dye in plastic containers. Unroll the roving—it should not be dripping wet—and place it on the plastic wrap. Depending upon the width of the roving, you may want to add several pieces (up to four) per piece of plastic wrap. Purchase a large syringe (no needle) or small plastic bottle with a tip. Add the dye to the syringe or bottle, and apply the dye to the roving. Add it in a manner that is pleasing to your eye. Shading can be subtle or bold. Various colorations and color values make for an interesting end result. Think of rainbows, Neapolitan ice cream, autumn pallets, and variances of light blue to indigo with splashes of purple. This is the place to let your inner designer come through! Be sure to make enough to spin up into a full project. There will be no recreating this exact yarn again, so you'll be crafting a one-time-only project. Of course, similar yarns can be reproduced, but there will always be slight variances (dye lots), which is what makes this project interesting.

Once the desired effect is achieved, remove the tape and roll up the roving in the plastic wrap, making a packet. It is okay if the ends don't meet in the middle. Steaming is required to set the dye. Using a kitchen canner, add enough water to cover the bottom of the kettle. Place the roving, still in the plastic wrap, in a rack, suspended above the hot water. Remember, you want steam, not boiling water. A rice cooker will work for this, too. Allow the roving to steam for about forty-five minutes. Make sure the kettle does not boil dry.

Remove the roving from the pot, being careful of the steam and hot wool. Allow it to cool to the point where the wool can be handled. Have a sink full of hot water with a squish of dish soap ready, and then unroll the roving (throw away the plastic) and place the wool into the hot water. Let it rest for about fifteen minutes. Squeeze out the excess water, and drain the water from the sink. Refill the sink with warm rinse water and add the roving one more time. After fifteen minutes, drain the sink and squeeze the excess water from the dyed fibers. Spin them out (no agitation) in the washing machine, and then hang them to dry. The effect before spinning is almost like that of a tied-dyed technique. Skeins of yarn may be dyed in the same manner.

Space Dyeing

Another interesting means of dyeing fiber is space dyeing. Recently I dyed roving in a crock pot. No kidding, a crock pot. Of course, you'll need a designated pot for this method, and it will only work for a small quantity of roving. Used cookers are available at yard sales and thrift stores.

To space dye in a crock pot, you'll need four quart-size canning jars, white vinegar to set the dye, and four ounces of wool/mohair blend roving. Choose any kind of dye, such as Kool-Aid, Easter egg dye, home dyeing preparations, or fiber dyes.

Begin by preparing the dye bath in the canning jars. Mix a different color in each jar according to the directions. Add 1/4 cup of white vinegar for Easter egg dye or Kool Aid only. Add enough water to fill the jar almost to the top. Place the jars in the crock pot.

Divide the roving into four parts; add one part to each jar, leaving a tail. Dip the tail from one jar into the jar next to it and submerge the tip in the dye. This connection will bring the dye from one jar to the other, subtly blending the fibers, one color to the next.

Fill the crock pot with water, almost to the top of the jars, and then turn it on low for three hours and add the lid. Turn the crock pot off after three hours, but let the fiber remain in the covered cooker for three more hours. Allow the fiber to cool, and then place it in a sink filled with warmish water to rinse. Rinse until the water runs clear. Spin the roving out in the washing machine, being careful not to agitate it. Hang the roving to dry, spin up the yarn, and glean the rewards.

Mohair and cashmere can create beautiful woven garments, and if you have access to a loom, you can experiment with spinning the fibers and creating luxurious garments. If you have your own fiber to work with, so much the better. You may also substitute these fibers as you spin, to create wonderful yarns for knitting or crocheting. For information on spinning, weaving, felting, and creating projects with goat fiber, please see *The Whole Goat Handbook*.

9 ★ GOAT MILK AND CHEESE

Reblochon

A variety of cheeses are made on Baetje Farm in Missouri from goats' milk.

Reprinted with the permission of Scribner Publishing Group from *Goat Song: A Seasonal Life, A Short History of Herding, and the Art of Making Cheese* by Brad Kessler.

Raw

Although we could sell our fresh raw milk directly from—and only from—our farmstead, selling our fresh raw-milk *cheese* was out of the question. No matter how healthy our goats, how virgin our pasture, how scrupulous our cheesemaking, if we sold or traded our chèvre we'd be breaking federal law. The US Food and Drug Administration law Title 21 CFR Part 133 prohibits the sale of raw-milk cheese that hasn't been aged at least sixty days. Our chèvre and mozzarella weren't aged at all.

To pasteurize milk the liquid is heated to at least 161 degrees Fahrenheit and held there for fifteen seconds. The heat kills most of the bacteria that cause illness—*Escherichia coli*, *Listeria monocytogenes*, *Salmonella*. Yet it also destroys the nutrients in the milk, its enzymes and vitamins—not to mention its flavor.

Milk comes out of a mammal alive with microorganisms. The microbes exist to nourish and help the survival of the mammal's offspring. Some organisms, such as the macrophages and T lymphocytes, aid the infant's immune system; others, such as lactoferrin and lysozyme, kill harmful bacteria. The enzymes—peroxidase, catalase, phosphatase, amylase, lipase, galactase—help digestion, while the oligosaccharides are indigestible and seem to exist solely to feed beneficial bacteria living inside the infant's stomach. All these compounds are found in raw milk—whether the milk of a cow, goat, horse, human, or whale. Pasteurization kills them all and turns the milk into a dead thing.

For thousands of years people believed fresh raw milk was a panacea. Hippocrates prescribed milk for people with tuberculosis. Arab physicians touted camel's milk. German and Russian physicians in the nineteenth century popularized the "milk cure," which was said to treat everything from liver disease to asthma. In the United States the Mayo Foundation (a forerunner to the clinic) promoted its own version of the milk cure, insisting that raw milk cured a plethora of ailments. What all these lactic enthusiasts shared was the belief in the power of unpasteurized milk from a *healthy animal fed what she was meant to eat*—namely grass.

Raw goat milk from healthy, pastured goats has long been considered a cure for various ailments.

Yet things didn't always work out so well for the cow. As a way of cleaning up the wastes from beer and whisky making (and turning an extra dime), American distillery owners in the nineteenth century crammed dairy cows into cellars and bricked enclosures and fed them hot, fermented distillery waste. The resulting milk, called swill or slop milk, was notably blue and often deadly and sold on the cheap to the poor. Some dairies added chalk or plaster of Paris to their milk. Once milk became transportable by train and then truck, milk traveled from farther afield, and city dwellers could no longer verify the cleanliness of the place their milk came from. Unsafe milk caused outbreaks of diphtheria, scarlet fever, typhoid fever, tuberculosis, brucellosis. It's no wonder that at the turn of the twentieth century the cry for clean milk was considered a moral cause.

Pasteurization was one method of assuring safe milk; another was strict inspection and certification of dairies. Each method had its advocates. But certification—a process of making sure the animals were healthy and the dairies spotless—lost out to the quicker fix: pasteurization. Pasteurization worked. People no longer died from drinking milk. Yet pasteurization often became an excuse for dairies to sell, not clean milk from healthy animals, but filthy milk from sick animals whose milk had been cooked clean of its impurities. Rather than rigorously certify raw-milk dairies—as is done in Europe today—it was less costly for the American dairy industry to simply zap their milk. Throughout the twentieth century, compulsory pasteurization laws in the United States expanded state by state, until it became nearly impossible for Americans to find anything but pasteurized—and effectively dead—milk.

Today in North America the typical industrial cow (organic or not) lives in a slurry of manure. She never walks on pasture or eats grass. She's fed what she was never intended to eat—high-energy grain. She will produce twice as much milk by eating grain instead of grass, but she'll become chronically sick from it. A diet of grain will lead to knotted guts, bloat, abscessed livers, ulcerated rumens, rotten hoofs, udder inflammation, shock, and eventually death. The average industrial milk cow lives less than five years. A cow allowed to eat grass on pasture can live fifteen or longer.

In a tiny operation like our own, there's no need to pasteurize the milk. We know beforehand if a doe is sick, and the quality of the milk is obvious because it sits right beneath our noses. Several states (such as

Vermont) recognize the relationship small farms have with their animals, and allow farmers to sell raw milk directly from their farmsteads. The farther milk has to travel from the teat, the greater the chances it can degrade. If our own raw milk was trucked out of the valley and traveled a few hours and sat in traffic on a city street and then on a shelf somewhere in a store, we couldn't vouch for its safety. Not because of what happened here *in situ*, but what might've happened once it left the valley. All this argues for a local source: an animal and a farmer one actually knows.

The debate over raw versus pasteurized gets a lot of people up in arms. Unpasteurized milk is the birthright of most Europeans, and when you tell someone from France that in the States you can't buy a fresh raw-milk cheese, they look at you as if you've just profaned the Madonna. It confirms everything they suspected about American culture: that there is none—especially when it comes to the cultures inherent in milk.

The simple truth is that you can't make a top-quality cheese from pasteurized milk. Pasteurization kills over 99 percent of the milk's bacteria, including all the good, but not necessarily the bad, bacteria. Pasteurization incinerates the building blocks of a good cheese—the *lactobacillus* indigenous to the milk. It also destroys the aromatic esters—the monoterpenes and sesquiterpenes—from the plants the animal's been eating, which give raw-milk cheese its unique herbal flavors, its *terrior*.

Raw milk is a living galaxy under a microscope. Commercial cheesemakers today go to great lengths trying to repopulate this milky way once it has been destroyed through pasteurization. But nothing concocted in a lab can replicate the diversity of microorganisms found in raw milk; and nothing can replace the natural antibiotic compounds—the lactogens—that inhibit the growth of harmful bacteria. Since pasteurization kills these protective inhibitors, pasteurized milk may be, ironically, *less safe* than raw during and after cheesemaking.

As advocates of the milk cure surmised long ago, there's something resident in clean raw milk that's good for humans. All mammals, after all, grow and become healthy on unprocessed raw milk. Scientists surveyed fifteen thousand children across Europe in 2007 and found that those who drank raw milk were practically free from all allergens. Those who drank pasteurized were not.

Pasteurization is necessary because of the risks inherent in distance: the farther milk travels from the goat, the more chances it has to become contaminated. If the milk comes from your own goats, you know exactly how clean and fresh it is.

I never used to think much about milk. I hardly touched the stuff. Pasteurized cow milk made me slightly ill. Aside from a splash in tea, I avoided dairy. I didn't much care for cheese, not because of the taste, but how it made me feel after. The goats changed all that. Goat milk is a lot closer to human milk than cow milk. Its fat globules are smaller and easier to digest and more resemble those of human milk. People drank goat milk for thousands of years before they ever tried the milk of a cow, which might explain why so many humans are still lactose intolerant to cow milk, and not to goat.

That first summer drinking unpasteurized goat milk I never felt healthier in my life. I didn't fall sick once, or come down with a cold, and the petty allergies that had plagued me for years disappeared. As Gandhi said: the goat proved a mother to me.

Making Cheese

If there is a dairy animal present, there will also be cheese. The cheesemaking techniques are similar everywhere, with variations on the central theme of milk, heat, agitation, and enzymes. In early times, before reliable refrigeration, cheesemaking was the only way to preserve milk for the inevitable dry spell. Dairy animals, being of seasonal lactation, do have a period of the year when they cease milk production. Aged cheese added a valuable measure of nutrition to what we can only assume was a sparse diet.

When modern-day science entered the picture, an understanding of the chemical process involved in cheesemaking developed. Amazingly, little has changed since the early times. First milk is warmed, and then a bacterial culture is added, followed by rennet, which is still produced from a calf's stomach. The culture acidifies the milk, and the rennet causes the milk to thicken, creating the gel that will become cheese. The rennet and culture continue to work during the aging process, bringing the cheese into maturity.

A simple chèvre, made from your own goats' milk, has a taste like nothing else.

Reprinted with permission from *Homemade Cheese* by Janet Hurst (Voyageur Press, 2011)

Milk: The Primary Ingredient

Small-scale cheesemaking is possible today both on a farm or in the middle of the city. Raw milk can often be purchased from local dairy farmers. Make sure the milk is clean and fresh before you make a purchase. Ask to taste it.

Off-tasting or sour-smelling milk will make off-tasting and sour-smelling cheese, so start off right! Laws governing raw milk sales vary from region to region. Check with the local health department or state department of agriculture to learn about the legality of purchasing raw milk.

Harmful bacteria in milk has a phenomenal self-replication rate, and bacteria counts increase every hour. It is easy to see how milk that has not been properly refrigerated can quickly become a problem. In his report "Hygiene and Food Safety in Cheesemaking" from the Vermont Institute of Artisan Cheese, Todd Prichard cites these statistics: "Temperature abuse is the #1 cause of food borne illness. Food must be moved through the Danger Zone as rapidly as possible. We must control the growth of unwanted bacteria or they will rapidly increase in numbers and potentially spoil the end product. Bacteria multiply exponentially (i.e., 1>2>4>8). It will only take 20 generations for one bacterium to become 1 million bacterium." Proper cooling and refrigeration of milk is essential. Safety first.

Cream-line milk is available in major grocery stores. The cream has not been separated from this milk, so a thick layer of real cream rests at the top of the bottle. This milk has been pasteurized and will work well for the small-scale cheesemaker. For our purposes, whole milk (milk that has not been separated) will be used.

To further explain the chemical processes of cheesemaking is to understand the composition of milk. Basically, milk is composed of water, lactose, fat, protein, minerals, and miscellaneous components, such as enzymes, vitamins, and somatic cells. The goal in cheesemaking is to isolate the solids in the milk, then to expel most of the moisture. The liquid removed during this process is whey. Whey is considered a waste product, except in the manufacture of ricotta or other whey-based cheese. It is also used within the health industry as a nutritional supplement. To make cheese, the cheesemaker brings milk to the temperature required to promote the growth of the bacteria that feed on lactose.

Do not purchase UHT milk for making cheese. UHT stands for ultra heat treated. Due to the high temperatures involved in the manufacture of a shelf-stable product, all bacteria contained in the milk is destroyed. Milk normally contains bacterial flora that can be enhanced by the addition of manmade cultures. However, when milk is exposed to the extremely high temperatures required for UHT, it is no longer suitable for cheesemaking. No bacteria, no cheese.

HOME MILK PASTEURIZATION

Pasteurizing milk at home on the stovetop is a simple process. An added bonus is that your milk won't have to stand up to shipping and prolonged storage, so you can pasteurize it safely using lower heat and taking less time than many industrial milk producers use, thus retaining the necessary bacteria for cheesemaking. All you need is a double boiler or two stainless-steel pots and a kitchen thermometer. Then just follow these simple steps:

1. Pour the milk into the smaller of the two pots, and place the small pot inside the larger pot, with 3 inches (7.6 cm) of water in the bottom.
2. Slowly heat the milk to 145°F (63°C) and hold the temperature there for 30 minutes. Stir the milk gently throughout the process to make sure it is evenly heated.
3. Remove the milk from heat and place it in a sink filled with ice, to bring the temperature down as quickly as possible. When the milk reaches 40°F (4°C), it is chilled and may be stored in the refrigerator until ready for use.

THE THREE COMMANDMENTS OF SANITATION

For the purposes of this book, I will assume these three things regarding sanitation:

1. The reader will supply a clean working environment. Hands and nails will be scrubbed thoroughly before any cheesemaking begins.
2. All equipment will be sanitized; this means pots, molds (for shaping cheese), spoons, ladles, cloths, knives, mats, and so on.
3. Milk purchased for cheesemaking will be from a reliable source and pasteurized according to directions. Raw milk products must be aged sixty days or more to be considered a safe food.

Sanitation

Cleanliness of utensils, work surfaces, and cooking pots is of utmost importance in cheesemaking. Bacterial contamination will occur if strict sanitation procedures are not developed and followed.

To sanitize equipment, fill your clean kitchen sink with tap water; add one cap of regular household chlorine bleach to sanitize utensils, pots, and cheese molds. Or take a tip from homebrewers of beer and wine, and use One-Step Sanitizer, an easy cleaner that requires no bleach or rinsing. Sanitize your equipment before you begin making cheese, and allow the equipment to air dry. Wash any cloth towels or cheesecloth in a mild bleach solution before and after use.

The average daily yield of a goat is a little less than one gallon per day.

Cheesemaking Techniques

Over time you will develop specific cheesemaking techniques.

Words not spoken or performed in the cheese room include beat, whip, mash, and chop. Cheesemaking is a gentle art, especially on the home-kitchen scale. Think Zen. Milk is fragile. If it is handled roughly, the fat cells, which are needed for the creation of cheese, will break down. So gentle handling of the milk is crucial to the cheesemaking process.

Here are some general tips to guide the new cheesemaker to success:

- Use one gallon of milk as a base line. In the beginning, do not increase the amount, just in case things do not go as planned. Pasteurize the milk, as explained.
- Set aside some uninterrupted time for the first venture into cheesemaking. Have all equipment ready and milk on hand. Cheesemaking requires patience, and attempts at shortcuts will lead to failures.
- Cultures are delicate beings. Always use a clean and dry spoon to retrieve the culture from its foil pouch. Add the specified amount to the warmed milk. Immediately close the pouch, clip it closed with a paper clip, and then place the foil pouch in a zip-top plastic bag. Refrigerate.
- If the make procedure calls for more than one culture, be careful not to cross contaminate one culture with another. Use a separate clean and dry measuring spoon for each type of culture.
- When heating milk for cheesemaking, do so with a low flame, being careful to avoid scorching.
- A timer is a cheesemaker's best friend. After you achieve a comfort level with the process, you will have other tasks to tend to instead of watching the pot. When that happens, set the timer. Time does fly, and it is easy to forget about the pot on the stove. The timer is a most valuable tool in the cheese kitchen.
- A good thermometer is one of the most helpful tools you can have. Temperature is one of the key components to good cheesemaking, so a thermometer is an invaluable and necessary addition to the basic equipment required for cheesemaking.
- Cheesecloth has come a long way. There is a synthetic blend perfect for draining curds. Traditional cheesecloth is often too thin to capture the curd, so the synthetic or a more traditional cloth, butter muslin, makes the best choice for draining. Nylon parachute fabric, usually available at fabric stores, also works wonderfully. It has the qualities of being porous enough to allow the whey to drain, yet captures even tiny bits of curd. Cheese-supply houses also offer draining bags, which provide an easy way to drain curds and whey.

Cheesemaking Step by Step

Pour the milk gently into your large cooking pot, and heat the milk slowly to 86°F (30°C) over medium heat.

Add the culture, and stir in the culture thoroughly using a top to bottom motion. Let sit for the time specified in the recipe.

Add the rennet. Some recipes will require you add the rennet drop by drop; others have you dilute the rennet in water and add the solution.

After adding the rennet to the milk, stir, top to bottom, for one minute.

Cover and allow the renneted milk to rest, undisturbed, for thirty minutes.

After slicing through the curd, go back with the knife and pick up the curd to see if there is a clean separation.

The curd after cutting. Note the curds are all about the same size. Let the curds rest for fifteen minutes after the cut is complete. This resting period allows the curds time to heal and toughen up.

While the cut curds are resting, prepare the cheese baskets by lining them with disposable cheesecloth. Place the baskets on a rack above a bowl or bucket that will catch the whey.

Gently ladle the curd into the lined baskets. The curds will compress as the whey is expelled, so expect the cheese to be about half the size of the original mold.

Fill the baskets to the top. Let the curds settle for two hours and then top off again. Allow the curds to drain for twelve hours. These curds will shrink to about half their original size.

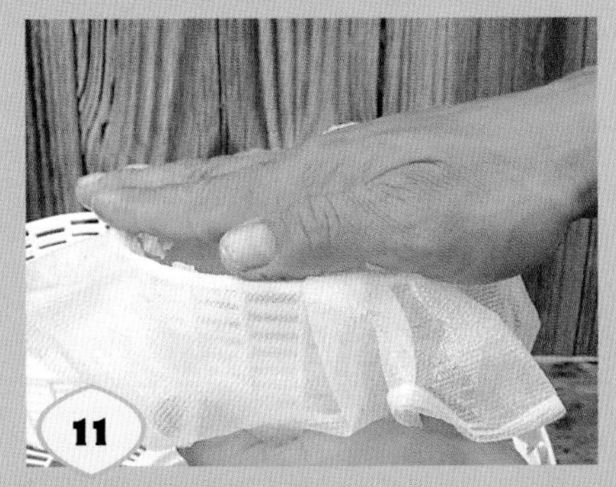

For a more even appearance, after six hours remove the cheese from the mold and flip it, the top becoming the bottom. Put the cheese back in the mold.

After twelve hours of draining, carefully take the cheese from the basket and place on the drying rack.

Salt the cheese to taste.

The classic French chèvre is one of the most basic cheeses. The word chèvre means "goat" in French.

Chèvre

Chèvre is only made from goat milk, so if you don't have goats, make an acquaintance with your local goat farmer. According to my French auntie, Elaine, the proper pronunciation is "chev."

This cheese is commonly produced in France by farmstead cheesemakers. It can be made with a minimum of skill, ingredients, and equipment, which makes it a perfect project for the beginning cheesemaker. A few purchases will be required to begin cheesemaking, so plan ahead to have the necessary equipment on hand.

Equipment

Slotted spoon
Thermometer
Ladle
String
Colander

Ingredients

1 gallon pasteurized goat milk
1/8 teaspoon Mesophilic DVI MA culture
2 drops of liquid rennet dissolved in 1/4 cup nonchlorinated water
1/2 to 1 teaspoon noniodized salt to taste
Optional: Herbs, such as fresh chives, lavender blossoms, or a blend, such as herbes de Provence; other ingredients, such as black pepper, green peppers, or olives.

Pour the goat milk into a cooking pot. Heat milk slowly to 86°F (30°C). Remove from heat.

Sprinkle the culture over the top of the milk and gently stir, making sure the culture is dissolved and well integrated into the milk. Allow this mixture to sit for about 45 minutes, so the culture has time to develop.

Add the rennet mixed in water and stir, coming up from the bottom of the pot, until the culture and rennet are well integrated into the milk. Let the mixture rest, covered with a cloth, in a warm place for 12 to 18 hours. The gel will thicken to the consistency of yogurt while it is resting.

When the gel has thickened, it is time to ladle the mass into a draining bag. Line a colander with the draining bag, cheesecloth, or muslin. Place the colander in the sink. With a slotted spoon, gently transfer the gel mass, now called the curd, into the lined colander. Keep ladling until all the curd is in the colander. The leftover liquid is the whey, which is a waste product. Once all the curd is in the colander, gather the draining bag and tie it with the string. Hang it over the sink, and the whey will drain, rapidly at first, then more slowly.

Two things are happening while the curd drains: Acid is developing, so the flavor of the cheese is coming to life. And the moisture ratio of liquid to solid is dropping; therefore, the consistency and the stability of the finished product are changing. Chèvre is meant to be soft, so the moisture level will remain high. But this high moisture makes chèvre less stable than other aged or hard cheeses, so it should be consumed within a few days of the make. (In the language of cheese, the process of creating the cheese is called "the make.")

Allow the curd to drain for about 12 hours. Then remove the curd from the bag, place it in a bowl, and work in the salt. Salting has a number of purposes in the cheesemaking process. It adds flavor, promotes the shedding of moisture, and retards bacteria growth. Salt can be added directly to the curd, used to develop the rind on the cheese with a direct rub, or added to water to create brine, which the cheese can be placed in.

Flavor with herbs or other ingredients. These ingredients can be added to the cheese to make a spread, or the cheese can be rolled into logs or rounds and then rolled in the herbs. Chèvre is somewhat bland, so it will take on the flavors of the condiments or herbs added to it.

To store, place in a covered dish. Best served at room temperature.

Raw-Milk Farm Cheese

This recipe works well with either goat or cow milk. When using raw milk, 60 days is the minimum time for aging a cheese before consumption, but it's not necessary to use raw milk.

Ingredients

2 gallons whole milk
½ teaspoon Mesophilic DVI MA culture
¼ teaspoon liquid rennet dissolved in
½ cup nonchlorinated water
2 tablespoons noniodized salt
(kosher salt is ideal)

Fill a large cooking pot with the milk. Warm the milk to 86°F (30°C). Sprinkle the culture over the milk and mix in, stirring top to bottom.

Add the rennet solution. Let the milk and rennet rest for about 20 minutes, undisturbed. Test for a clean break.

When the clean break is achieved, cut into ½-inch (13-mm) cubes by cutting one way and then the other, to make a pattern of squares on top, and then by reaching in to cut across from the top to the bottom of the pot.

Bathe the curds for 2 minutes by moving them around in the whey gently and slowly with your hands. This helps the curds toughen a bit.

Set your cooking pot in a metal dishpan of water and place the assembly back on the stove. Keep it covered with a cloth. Slowly warm to 102°F (38.8°C), stirring occasionally. Remove the dishpan-and-pot assembly from the heat and let it sit an hour, stirring occasionally and gently.

Line a colander with cheesecloth, fastening the edges down with clothespins if you'd like to hold it in place. Gently ladle the curds into the colander and drain the whey.

Add the salt and mix it into the curd with your hands.

Tie the ends of cheesecloth together to make a bag. Hang it where it can drain. Let drain for about 2 hours in a warm place.

Leaving the curds in the bag, place the bag on a rack, such as a rack used for cooling cookies. On top of the bag, place a plate, then top with a weight, such as a plastic jar filled with water, a brick, or some other heavy object weighing about 5 pounds. Allow to press overnight.

Remove from cheesecloth, and let the cheese air dry for 2 or 3 days, until it forms a rind.

Wax the cheese and store in a cool, dry place for 60 days or more.

Cream Cheese

Cream cheese has become a commodity product. Bring it back to life and make your own. It is a simple task, and nothing surpasses the announcement of "I made the cheese myself."

Ingredients

1 gallon pasteurized whole goat or cow milk
¼ teaspoon Mesophilic DVI MA culture
3 drops of liquid rennet in ⅛ cup of nonchlorinated water (measure out 2 teaspoons of this mixture and discard the rest)
noniodized salt, to taste

In a large cooking pot, warm the milk to 86°F (30°C). Add the culture, then the rennet solution (remember, only 2 teaspoons of the dilution). Cover the pot and allow it to sit at room temperature (70°F or 21°C) for 12 to 18 hours. It will have the appearance of yogurt.

Line a colander with cheesecloth, doubling the cloth to catch the curd, or use butter muslin (a finer cheesecloth). Ladle the curd into the colander and allow it to drain for about 12 hours. The best way to drain it is to hang the bag over the sink to allow the weight of the cheese to compress the curd.

After 12 hours, remove the cheese from the bag and work in the salt.

Refrigerate the cheese. When it is well chilled, you may make it into logs or blocks to resemble traditional cream cheese. Eat and enjoy.

Kefir

Kefir is a fermented type of milk with a consistency similar to grocery-store yogurt drinks. It's a common drink throughout the Middle East.

Ingredients

1 tablespoon kefir grains
1 quart whole or 2 percent milk

Put kefir grains in a glass jar and fill the jar almost full with the whole milk. Cover with a clean cloth and set aside on your kitchen counter. Wait 1 to 2 days, stirring periodically with a plastic spoon. (Using plastic is particularly important, since metal appears to damage the cultures.) When the milk is thick, strain out the kefir with a plastic strainer (being careful to keep the grains intact). The milk that was strained is ready for use. Rinse the used grains, refill the jar with fresh milk and restart the process.

Next time you wish to make kefir, you can use these same grains, as they will continue to remain active. Just pour them into a glass jar, cover them with water, seal the jar, and refrigerate it.

Yogurt

Yogurt made from goat milk is tangy and pleasantly acidic. Often those who are allergic to cow milk can tolerate goat-milk products.

Ingredients

1 gallon fresh, unpasteurized goat or cow milk

1 tablespoon plain yogurt with active cultures or 1 packet freeze-dried culture containing lactobacillus

Heat the milk to 108°F (42.2°C). Add the plain yogurt or freeze-dried cultures. Make sure to use yogurt from a new cup and a clean spoon to add the yogurt.

Incubate the milk mixture at 104 to 108°F (40 to 42.2°C). To do this, you can use a home yogurt maker or an incubation device of your own. Some people use a heating pad wrapped around a jar; they put the jar in the oven on low, or place the jar in a crockpot. Whatever you use, experiment with water and a thermometer before you actually make the yogurt, to be sure you can hold the milk at the required temperature. Incubate the milk for 6 to 8 hours, depending on your taste.

When finished incubating, chill the yogurt before eating it, being careful not to agitate or move the yogurt much until it is well chilled.

Goat milk makes a deliciously tangy yogurt.

Goat-milk yogurt will not get as thick as cow-milk yogurt. Many commercial cow-milk yogurts add powdered milk as a thickener. You can also use a small amount of tapioca, which is a natural thickener from the cassava root.

Lebneh

Lebneh yogurt cheese is widely used in the Middle East and Greece—a fact that results in several spellings of its name, including lebanah and labanah.

This cheese is the same consistency of cream cheese. It is also easy to make. Simply allow the liquid to drain from the solids. Yogurt funnels are available for this very purpose.

To make yogurt cheese, place 2 cups of plain goat- or cow-milk yogurt in a colander lined with three layers of moistened cheesecloth. Bring the corners of the cheesecloth together to form a bag, which can then be drained over the sink. Let the yogurt drain for 8 to 16 hours. Stir occasionally, scraping the cheese away from the cheesecloth to allow better draining. The longer the yogurt drains the thicker and more tart the yogurt cheese will be.

Source: Jennifer Bice of Redwood Hill Farm and Creamery

RESOURCES

Books on Goatkeeping

Amundson, Carol A. *How to Raise Goats.* Voyageur Press, 2007, 2013.

Drummond, Susan. *Angora Goats the Northern Way*, 5th Ed. Stoney Lonesome Farm, 2005.

Hurst, Janet. *The Whole Goat Handbook.* Voyageur Press, 2013.

Zweede-Tucker, Yvonne. *The Meat Goat Handbook: Raising Goats for Food, Profit, and Fun.* Voyageur Press, 2012.

Books on Cheesemaking

Chadwick, Janet. *How to Live on Almost Nothing and Have Plenty.* Alfred A. Knopf, 1982; republished by CreateSpace Independent Publishing Platform, 2011 (with the subtitle *A Practical Introduction to Small-Scale Sufficient Country Living*).

Hooper, Allison. *In a Cheesemaker's Kitchen: Celebrating 25 Years of Artisanal Cheesemaking from Vermont Butter and Cheese Company.* Countryman Press, 2009.

Hurst, Janet. *Homemade Cheese: Recipes for 50 Cheeses from Artisan Cheesemakers.* Voyageur Press, 2011.

Morris, Margaret. *The Cheesemaker's Manual.* Glengarry Cheesemaking, Inc., 2003. Out of print.

Roberts, Jeff. *Atlas of American Artisan Cheese.* Chelsea Green Publishing, 2007.

Toth, Mary Jane. *Goats Produce, Too. Vol. II.* N.p.: n.p., 2007. (This self-published book is available through the cheesemaking supply sources listed.)

Books on Felting and Weaving

Hoerner, Nancy et al. *Easy Needle Felting.* Sterling, 2008.

Horvath, Marie-Noelle. *Little Felted Animals: Create 16 Irresistible Creatures with Simple Needle-Felting Techniques.* Potter Craft, 2008.

Matthiessen, Barbara. *Small Loom & Freeform Weaving: 5 Ways to Weave.* Creative Publishing International, 2008.

Rainey, Sarita R. *Weaving Without a Loom, 2nd Ed.* Davis, 2008.

Warner Dendel, Esther. *Needleweaving . . . Easy as Embroidery.* Doubleday, 1976.

Supplies

Caprine Supply
Dairy-goat supplies, cheesemaking supplies
www.caprinesupply.com
P.O. Box Y
De Soto, KS 66018
Order line: 1-800-646-7736

Dairy Connection
Cultures, molds, equipment
Jeff Meier or Cathy Potter
www.GetCulture.com
501 Tasman Street, Suite B
Madison, WI 53714
608-268-0462

Glengarry Cheesemaking and Dairy Supply
Margaret Morris
www.glengarrycheesemaking.on.ca
Canada: 5926 Highway #34, RR #1, Lancaster, Ontario, Canada K0C 1N0
USA: P.O. Box 92, Massena, NY 13662, USA
1-888-816-0903 or 613-347-1141

Hoegger Supply Company
Cheesemaking and goat supplies
www.hoeggerfarmyard.com
200 Providence Road
Fayetteville, GA 30215
1-800-221-4628

Leeners
Cheesemaking kits and supplies
www.leeners.com/cheese/store
9293 Olde Eight Road
Northfield, OH 44067
1-800-543-3697

Lehman's
Cheesemaking kits and other goat-related products
www.Lehmans.com
4779 Kidron Road
Dalton, OH 44618
1-888-438-5346

New England Cheesemaking Supply
Ricki Carroll
www.cheesemaking.com
54B Whatley Road
South Deerfield, MA 01373
413-397-2012

Jack Schmidling
Cheese presses, cheese recipes
http://schmidling.com/cres.htm
18016 Church Road
Marengo, IL 60152

Soapmaking Supplies

Bramble Berry
www.brambleberry.com
2138 Humboldt Street
Bellingham, WA 98225
360-734-8278

Technical Resources

ATTRA (National Sustainable Agriculture Information Service)
www.attra.org
P.O. Box 3838
Butte, MT 59702

CheezSorce
Cheesemaking business consultation.
www.cheezsorce.com
4207 McCausland Avenue
St. Louis, MO 63109
314-517-4397

David Fisher
The online guru for candle and soap makers.
www.candleandsoap.about.com

SARE (Sustainable Agriculture Research and Information)
www.sare.org
1122 Patapsco Building
University of Maryland
College Park, MD 20742

University of Minnesota Extension
Research-based information from a network of university experts.
www.extension.org

Research-Based Goat Programs

Cornell University
ansci.cornell.edu/goats/

Langston University
www.luresext.edu/GOATS/index.htm

Lincoln University Cooperative Extension (LUCE) and the Innovative Small Farmer's Outreach Program (ISFOP)
www.lincoln.edu/web/programs-and-projects/innovative-small-farmers-outreach-program

Texas A&M
animalscience.tamu.edu/about/facilities/sheep-goat/

Magazines

Countryside (www.countrysidemag.com)
A homesteader's delight.

Culture (www.culturecheesemag.com)
A magazine made for cheese lovers.

Dairy Goat Journal (www.dairygoatjournal.com)
A great resource for goat lovers.

Hobby Farms (www.hobbyfarms.com)
Detailed information on starting your own farm.

Mary Jane's Farm (www.maryjanesfarm.org)
A farm girl after my own heart.

Mother Earth News (www.motherearthnews.com)
Reading this magazine got me started down the goat path.

Sheep! (www.sheepmagazine.com)
By the publishers of the *Dairy Goat Journal* and *Countryside*.

Small Farmers' Journal (www.smallfarmersjournal.com)
Great stories of farm life.

Small Farm Today (www.smallfarmtoday.com)
A Missouri-based publication with great information on farming.

Websites

A Campaign for Real Milk (www.realmilk.com)
Farm sources where you can purchase milk directly.

American Dairy Goat Association (www.adga.org)

Fias Co Farm (www.fiascofarm.com)
A wealth of information on goats and cheesemaking.

Janet Hurst (www.cheesewriter.com)
Information on cheesemaking classes, blogs, and more.

The Livestock Conservancy (www.livestockconservancy.org)

Local Harvest (www.localharvest.org)
Sources from which to purchase milk, cheese, and other farm offerings.

SmallDairy.com (www.smalldairy.info)
A great resource for those interested in small dairies and in milk and cheese production.

Other Resources

American Cheese Society
Find cheesemakers in your area; guild listings, education, and more.
www.cheesesociety.org
2696 S. Colorado Boulevard, Suite 570
Denver, CO 80222
720-328-2788

California Polytechnic State University (Cal Poly)
www.calpoly.edu
San Luis Obispo, CA 93407
805-756-1111

The Goat Blaaag
A goat-related blog
www.theblaaag.com

Vermont Institute of Artisan Cheese
Technical consultation for Vermont and New England cheese makers
www.uvm.edu
University of Vermont
Burlington, VT 05405
802-656-3131

Washington State University Creamery
creamery.wsu.edu
P.O. Box 641122
Pullman, WA 99164-1122
800-457-5442

University of Wisconsin, River Falls
www.uwrf.edu
410 S. Third Street
River Falls WI 54022-5001
715-425-3911

INDEX

Items in bold face indicate a recipe. Page numbers in italics indicate an item appearing in a photograph, illustration, or caption.

Photo credits

Front and back matter: frontis Cody Wheeler/Shutterstock; title page Trinity Mirror/Mirrorpix/Alamy; page 5 Dennis van de Water/ Shutterstock; page 109 Dietmar Heinz© Picture Press/Alamy; page 111 FPG/Hulton Archive/Getty Images • **Introduction:** page 6 Janet Hurst (inset), Petar Paunchev/Shutterstock; page 7 Jodie Bell/Getty Images; page 8 Wolfgang Zwanzger/Shutterstock (top), Janet Hurst; page 9 Janet Hurst (top), AnetaPics/Shutterstock • **Chapter 1:** page 10 Library of Congress; page 11 © Doug Steley C/Alamy (top), SSPL via Getty Images; page 12 (top to bottom) mountainpix/Shutterstock, Prisma Archivo/Alamy, © Image Asset Management Ltd. /Alamy, Prisma Archivo/Alamy, SSPL/Getty Images; page 13 Library of Congress; page 14 Popperfoto/Getty Images; page 15 Hulton Archive/ Getty Images; page 16 Kean Collection/Getty Images; page 17 Library of Congress • **Chapter 2:** page 18 Im Perfect Lazybones/Shutterstock (left), © Emilio Ereza/Alamy; page 19 rscreativeworks/ Shutterstock; page 21 (clockwise from top left) Gala_ Kan/Shutterstock, Kerry Sherck © Aurora Photos/ Alamy, Ginger Livingston Sanders/Shutterstock, Tom Curtis/Shutterstock, Janet Hurst, © Mark Scheuern/Alamy; page 22 Ann E. Parry/Alamy (top), Robbie Taylor/Shutterstock (left); page 23 col/Shutterstock (left), Sue McDonald/ Shutterstock (top), TFoxFoto/Shutterstock; page 24 Lorri Carter/Shutterstock; page 25 © amanda hughes / Alamy (top), Teerinvata/Shutterstock • **Chapter 3:** page 26 Library of Congress (left), Art Directors & TRIP/Alamy; page 27 Heritage Image Partnership Ltd/Alamy; page 28 Lordprice Collection/Alamy (top), Lebrecht Music and Arts Photo Library/Alamy; page 29 (right) The British Library/Robana via Getty Images; Page 30 The Art Gallery Collection/Alamy (top), Mary Evans Picture Library/Alamy • **Chapter 4:** page 32 Michel Sima/RDA/Getty Images; page 33 DEA / G. DAGLI ORTI/De Agostini/Getty Images (inset), De Agostini/Getty Images; page 34 (clockwise from upper left) DEA/A.Dagli orti/De Agostini/ Getty Images, DeAgostini/Getty Images, The Walters Art Museum, DEA/G. NIMATALLAH/De Agostini/Getty Images; page 35 George Eastman House/Getty Images (top), Lee Boltin/Time & Life Pictures/ Getty Images; page 36 SuperStock/Alamy (top), Marco Secchi/Alamy (left), Faiz Balabil/Alamy; page 37 William Vanderson/Fox photos/ Getty Images; page 38 Bobkeenan Photography/Shutterstock (top); page 44 Photos 12/Alamy (inset), Universal Pictures/Getty Images; page 45 Hulton Archive/Getty Images (top), Buyenlarge/ Getty Images (left), Albert Harlingue/Roger Viollet/Getty Images; page 46 Moviepix/Getty Images (top), Murray Garrett/Getty Images (left), Moviestore collection Ltd/Alamy, Universal/courtesy Everett Collection; page 47 Moviepix/Getty Images (top), Photos 12/Alamy (middle and bottom); page 48 AP Photo/M. Spencer Green (top), Kim Karpeles/Alamy (bottom right), David Lee / Shutterstock.com (bottom left); page 49 maxhphoto / Shutterstock.com; page 50 Trinity Mirror/ Mirrorpix/Alamy; page 51 Library of Congress (top), INTERFOTO/ Alamy (bottom) • **Chapter 5:** page 56 Aurora Photos/Alamy; page 57 AP Photo/East Valley Tribune, Tim Hacker; page 58 Linda Davidson/ The Washington Post via Getty Images (both); page 59 AP Photo/ Patrick Semansky (top), AP Photo/Rick Bowmer; page 60 Library of Congress; page 61 Transcendental Graphics/Getty Images (top), Getty Images; page 62 Aurora Photos/Alamy (both); page 63 Tim Panell/Mint Images/Getty Images • **Chapter 6:** page 64 P.Uzunova/Shutterstock; page 65 oorka/Shutterstock; page 66 George Dolgikh/Shutterstock; page 67 gnomeandi/Shutterstock; page 68 Arena Photo UK/ Shutterstock (top), Stokkete/Shutterstock; page 69 Alekuwka/Shutterstock • **Chapter 7:** page 70 Im Perfect Lazybones/Shutterstock; page 71 © WaterFrame / Alamy; page 72 © Johner Images / Alamy; page 73 WDG Photo/Shutterstock (top), Square Dog Photography/Getty Images (left); page 74 Andrew Roland/Shutterstock; page 75 (clockwise from top left) Eric Isselee/Shutterstock, Four Oaks/ Shutterstock, Photoman29/Shutterstock, Julia Remezova/ Shutterstock; page 76 makspogonii/Shutterstock (top), Sean Gallup/ Getty Images; page 77 © Arco Images GmbH /Alamy; page 78 Andrew Roland/Shutterstock (right), Mircea BEZERGHEANU/Shutterstock; page 79 Rashid Valitov/Shutterstock; page 80 Anna Kucherova/ Shutterstock (top), dezi/Shutterstock; page 81 James Hager/Robert Harding World Imagery/Getty Images (top), © Hemis/Alamy; page 82 makspogonii/Shutterstock (bottom), Edouard Boubat Gamma-Rapho via Getty Images (top); page 83 LehaKoK/Shutterstock (top), Trinity Mirror/Mirrorpix/Alamy (bottom left), Paul Liebhardt /Alamy (right) • **Chapter 8:** page 84 Oliver Hoffmann/Shutterstock; page 86 Four Oaks/Shutterstock (top), Janet Hurst; page 87 Janet Hurst (top), v.schlichting/Shutterstock; page 88 Janet Hurst; page 89 Janet Hurst; page 90 Janet Hurst (both) • **Chapter 9:** all photos by Janet Hurst

More extraordinary books from Voyageur Press

How to Raise Goats
ISBN 978-0-7603-4378-4

The Whole Goat Handbook
ISBN 978-0-7603-4236-7

The Meat Goat Handbook
ISBN 978-0-7603-4042-4

Homemade Cheese
ISBN 978-0-7603-3848-3

How To Raise Chickens
ISBN 978-0-7603-4377-7

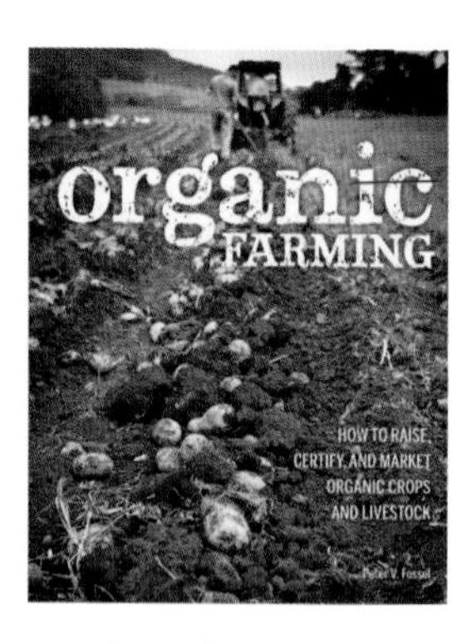

Organic Farming
ISBN 978-0-7603-4571-9

The Edible Landscape
ISBN 978-0-7603-4139-1